NATURAL DISASTER RESEARCH, PREDICTION AND MITIGATION

CYCLONIC AND GEOMAGNETIC STORMS

PREDICTING FACTORS, FORMATION AND ENVIRONMENTAL IMPACTS

NATURAL DISASTER RESEARCH, PREDICTION AND MITIGATION

Additional books in this series can be found on Nova's website under the Series tab.

Additional e-books in this series can be found on Nova's website under the e-book tab.

NATURAL DISASTER RESEARCH, PREDICTION AND MITIGATION

CYCLONIC AND GEOMAGNETIC STORMS

PREDICTING FACTORS, FORMATION AND ENVIRONMENTAL IMPACTS

VICTORIA P. BANKS
EDITOR

nova publishers
New York

NOTICE TO THE READER

The Publisher has taken reasonable care in the preparation of this book, but makes no expressed or implied warranty of any kind and assumes no responsibility for any errors or omissions. No liability is assumed for incidental or consequential damages in connection with or arising out of information contained in this book. The Publisher shall not be liable for any special, consequential, or exemplary damages resulting, in whole or in part, from the readers' use of, or reliance upon, this material. Any parts of this book based on government reports are so indicated and copyright is claimed for those parts to the extent applicable to compilations of such works.

Independent verification should be sought for any data, advice or recommendations contained in this book. In addition, no responsibility is assumed by the publisher for any injury and/or damage to persons or property arising from any methods, products, instructions, ideas or otherwise contained in this publication.

This publication is designed to provide accurate and authoritative information with regard to the subject matter covered herein. It is sold with the clear understanding that the Publisher is not engaged in rendering legal or any other professional services. If legal or any other expert assistance is required, the services of a competent person should be sought. FROM A DECLARATION OF PARTICIPANTS JOINTLY ADOPTED BY A COMMITTEE OF THE AMERICAN BAR ASSOCIATION AND A COMMITTEE OF PUBLISHERS.

Additional color graphics may be available in the e-book version of this book.

Library of Congress Cataloging-in-Publication Data

Cyclonic and geomagnetic storms : predicting factors, formation, and environmental impacts / Victoria P. Banks, editor.
pages cm. -- (Natural disaster research, prediction, and mitigation)
Includes index.
ISBN 978-1-63482-360-9 (hardcover)
1. Magnetic storms. 2. Cyclones. 3. Storms. I. Banks, Victoria P., editor.
QC835.C93 2015
551.51'4--dc23

2015010474

Published by Nova Science Publishers, Inc. † New York

CONTENTS

Preface

Solar activity causes changes in the magnetic field on the surface of the Earth. If these changes are insignificant, they do not affect life on Earth. However, rapid variations of the geomagnetic field, the so-called magnetic storms, which have the character of shock waves, may adversely affect living organisms (including humans), and the technical systems that people use. The book presents a conceptual design that allows prediction of electrical power system emergency conditions caused by magnetic storms. Finally, the authors' present a comprehensive analysis of perturbations in the low- and mid-latitude ionosphere during moderate recurrent geomagnetic storms.

Chapter 1 – The influence of variations of the geomagnetic field on electrification system is investigated with the aim to evaluate its negative effects on the operation of various electrical devices, such as systems for the transmission of electrical energy. An algorithm for calculating the currents that can be induced in the power lines and can be lead to the violation of stability of the system is derived. The algorithm is demonstrated on an illustrative example.

Chapter 2 – The spectrum of phenomena produced by geomagnetic storms in the near Earth's environment is gradually widened as observational data increase. One of them is significant enhancements of energetic electrons (>10 keV) revealed recently in the forbidden zone below the radiation belt. A comprehensive analysis of the forbidden electron enhancements during major and recurrent magnetic storms in solar cycles 23 and 24 was conducted on the base of experimental data acquired from low-orbiting satellites from 1998 to 2014. A possible mechanism of the phenomenon is discussed. Knowledge of where and when such environmental events will occur can be used for both space physics and applied researches. We focus on the following radiation aspects of the electron enhancement problem with a short review for each item: 1) ionization impact of 10-30 keV electrons in the ionosphere; 2) X-ray and gamma-ray contaminations by 100-300 keV electrons to astrophysical observation on board low-orbiting satellites; 3) possible effects of energetic electrons in satellite anomalies attributed to low-orbit radiation environment.

Chapter 3 – The authors present a comprehensive analysis of perturbations in the low- and mid-latitude ionosphere during moderate recurrent geomagnetic storms (RGSs) in 2007 – 2008 solar minimum. It is shown that during so-called recurrent ionospheric storms (RIS), the heliospheric – magnetospheric – ionospheric - thermospheric coupling remains highly efficient even during solar minimum. Analysis of global ionospheric maps of vertical total electron content (VTEC) derived on the base of GPS network showed that VTEC has a tendency to decrease and negative ionospheric storms occur right before the onset and on the late recovery phase of RGS. A positive ionospheric storm is developed 2 to 4 days after, i.e., on the recovery phase of RGS produced by high-intensity long-duration continuous auroral activity. The ionospheric response to heliospheric and geomagnetic drivers reveals prominent longitudinal and local time variations. Such effects as penetrating electric fields of interplanetary and magnetospheric origin, disturbance dynamo electric fields, neutral winds and thermospheric chemical composition changes are considered as possible mechanisms responsible for the observed recurrent ionospheric storms.

In: Cyclonic and Geomagnetic Storms
Editor: Victoria P. Banks
ISBN: 978-1-63482-360-9

Chapter 1

THE EFFECT OF GEOMAGNETIC STORMS ON THE ELECTRIFICATION SYSTEM

Daniel Mayer*
University of West Bohemia in Pilsen,
Pilsen (Czech Republic)

ABSTRACT

The influence of variations of the geomagnetic field on electrification system is investigated with the aim to evaluate its negative effects on the operation of various electrical devices, such as systems for the transmission of electrical energy. An algorithm for calculating the currents that can be induced in the power lines and can be lead to the violation of stability of the system is derived. The algorithm is demonstrated on an illustrative example.

Keywords: Geomagnetic storms, geomagnetic activity, collapse the electrical systems, matrix methods of electrical circuit analysis

1. INTRODUCTION

Solar activity causes changes (variation) of the magnetic field on the surface of the Earth. If these changes are insignificant, they does not affect life

* Corresponding author: mayer@kte.zcu.cz.

on Earth, nor the technical equipment. Exceptionally, however, a huge and rapid variations of the geomagnetic field, the so-called *magnetic storms*, which have the character of shock waves, may adversely affect living organisms (including humans), and the technical systems that people use. The more sophisticated technology humanity uses, the more is this technique vulnerable and thus the threat of strong magnetic storms becomes actual. Magnetic storms can cause serious damage to technical systems. They reduce the quality of the radio connections, may damage the cable telecommunication networks and satellite communications systems and thus block the data transmission, can interfere with navigational systems and thus jeopardizing the safety of air transport. Induced currents accelerate corrosion of pipelines transporting oil and gas, in railway track network where may disrupt railway signaling devices and, in particular, threaten the stability of large electrification systems, which can cause widespread blackouts and in limit case can even cause a total collapse of the system and global chaos, with disastrous consequences. There are even pessimistic forecast of extremely strong magnetic storms, according to which even the modern civilization may be threatened.

Geomagnetic variations cannot be avoided, but they can be (with certain approximation) predicted in the short-term. Geomagnetic activity is monitored by the global network of regional centers. Geomagnetic measurements are carried out continuously, at the interval of one minute, or even every second, with the accuracy of 0,1 [nT] and the results are forwarded to the World Data Center in Boulder (USA) and to the International Centre in Edinburgh (Scotland), Figure 1 (see [1]).

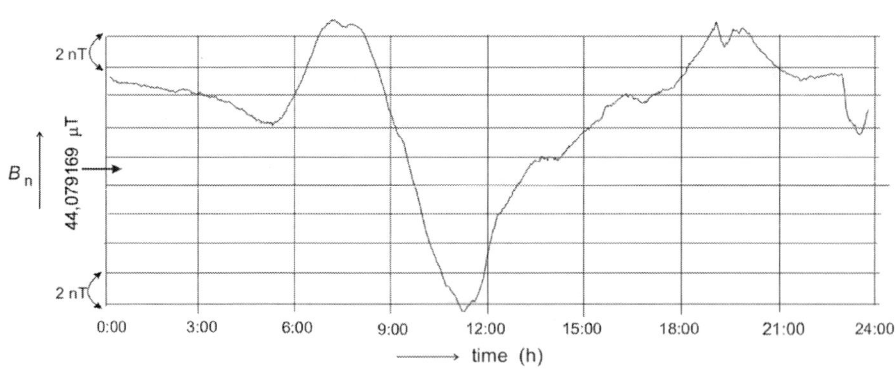

Figure 1. The time functions of the vertical component of the geomagnetic field during March 22, 2013, according to the measurement of the certified geomagnetic observatory in Budkov near Prachatice (Czech Republic) [1].

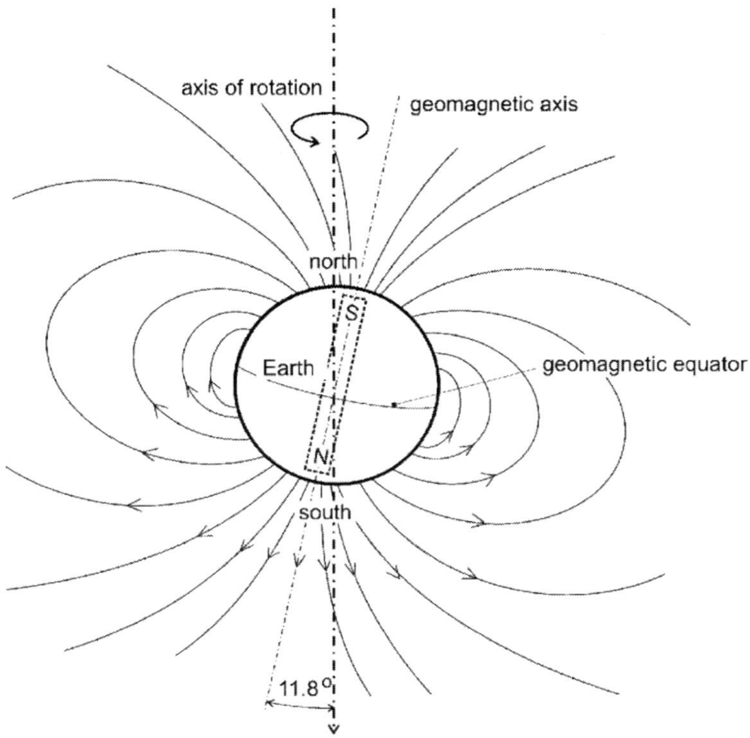

Figure 2. Geomagnetic field not disrupted by the solar wind.

This paper presents a conceptual design that allows prediction of electrical power system emergency conditions caused by magnetic storms.

2. The Origin of Geomagnetism and Geomagnetic Storms

Internal geomagnetic field. More than 400 years ago, W. Gilbert on the basis of experiments believed that there is a permanent magnet at the center of the Earth – a magnetic dipole – which generates a magnetic field on the surface of the Earth. There is the south magnetic pole in the northern hemisphere and the axis of the magnetic dipole is diverted by 11.8° from the Earth's axis (Figure 2). C. F. Gauss performed the mathematical solution of the geomagnetic field for the dipole configuration in 1828, using the spherical harmonic analysis. C. F. Gauss also developed an absolute method of measuring the geomagnetic field and together with A. Humboldt and E. Weber established a worldwide network of observatories for continuous measurement of the geomagnetic field. The first geomagnetic observatory in Central and Eastern Europe was founded by K. Kreil in Prague in 1839. Measurements of the geomagnetic field and plotting global magnetic maps in different times (epochs) showed that the geomagnetic field is variable not only with a place on Earth, but it is non-stationary, it changes with time – we call this phenomena as *secular variations*. These are the changes caused by physical processes in the Earth's core. These changes are very slow, they are detected on the scale of decades, and are completely irrelevant for our problem of geomagnetic field effects on the electricity system.

Gilbert's idea about the nature of geomagnetism with the permanent dipole magnet was abandoned after finding that the Earth's core temperature is above the Curie temperature; the ferromagnetic magnetic domains are thus randomly oriented. For example, at the depth of 100 km below the surface, the temperature is in the order of 1000 °C, which is significantly higher than the Curie temperature of the ferromagnetic material. Earth's magnetic field, which is variable with time and space on Earth, was explained by various hypotheses that have never been confirmed. For example Servington in r. 1732 theorized that complicated the magnetic field of the Earth is caused by the irregularities in the surface of the iron core of the Earth. In the first period of the last century, a model was designed to explain the geomagnetism – it was called *geodynamo*. Its physical nature is the rotational flow liquid of the outer core of the Earth, which is composed mainly of iron and nickel. Since it is an electrically conductive environment that lies in a magnetic field (in a weak magnetic field such as the magnetic field of the Sun), very high values of electric current are induced (about 10^9 A), which generates the geomagnetic field. The model of geodynamo and his subsequent mathematical description underwent a complex evolution, in which many leading astrophysicists participated. It was soon shown that the geodynamo simply could not be represented with an axially symmetrical, stationary flow, as initially anticipated. The most accurate model, developed by G. A. Glatzmaier and P. H. Roberts [2], has a much more complex configuration: it is a three-dimensional, turbulent, unsteady magnetohydrodynamic process, in which the applied laws of electromagnetic induction, the laws of fluid flow, electromagnetic and gravitational forces, Coriolis force, temperature convection and electric currents in the ionosphere are employed. Although present time computers allow only an approximate solution of this complex dynamic process, computer simulations can explain secular variation and even inversion (i.e., change of polarity) of geomagnetic field in the distant geological past of the Earth. (The development of the theory of geodynamo is described e.g., in the work of Chicago astrophysicist professor E. N. Parker [3, 4].)

The results obtained by computer simulations are compared with measured values obtained by geomagnetic observatories or by satellites. Archeological studies based on the determination of remanent magnetism burned archaeological objects of known age inform about secular variations of the geomagnetic field in the past centuries or millennia. For example, the samples derived from prehistoric kilns or fireplaces can be used to determine direction and magnitude of the magnetic intensity of the geomagnetic field at the time of firing the examined object. The information about the geomagnetic field of the Earth before tens of thousands to millions of years is obtained from paleomagnetic researches of remanent magnetization of quarternary volcanic and sedimentary rocks based on the decay of atmospheric carbon ^{14}C. The information on the geomagnetic poles wandering around the geographic poles in various geological epochs and then on multiple geomagnetic polarity inversion are acquired in the same way (for more details see [5]).

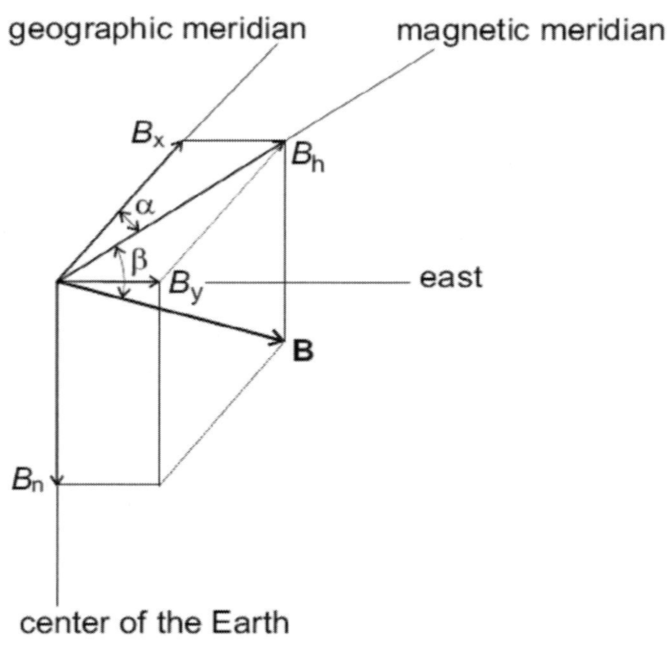

Figure 3. Vector of the geomagnetic filed and its components: $\boldsymbol{B} = \mathbf{i}\,B_x + \mathbf{j}\,B_y + \mathbf{k}\,B_n$.

Geomagnetic field generates the Earth's *magnetosphere*. The vector of magnetic flux density $\boldsymbol{B}$ has components B_x and B_y lying on the surface of the Earth (their resultant is the horizontal component B_h) and the vertical component B_n, which has the direction of the normal to the surface of the Earth (Figure 3). In geophysics the vector flux density $\boldsymbol{B}$ is usually determined from angles α (magnetic declination) and β (magnetic inclination) and from the normal component B_n. The magnitude of the magnetic field vector on the surface of the Earth is $B \sim 60\ \mu T$ at the poles, $B \sim 30\ \mu T$ at the magnetic equator and in the Central Europe is B ~ 44 μT.

External geomagnetic field. Electric currents in the ionosphere and the magnetic field of the Sun and other planets contribute to the magnetic field of the Earth. Over a hundred years ago it was found that there are fairly *rapid variations* of geomagnetic field that correlate with solar activity and show some periodicity – we are talking about daily, annual, eleven year and lunar variations. Variations, which originate from solar flares, when there is a surge of coronal mass containing electrons and high-energy protons and heavier ions, which then spread with great speed into interplanetary space, are crucial for our problem. This product of solar activity is known as a CMA (Coronal Mass Ejection) and its flow is called *the solar wind*. The delay between the occurrence of solar flares and geomagnetic field variations usually lasts one to two days. The process of flow of the solar wind in the interplanetary space undertook the marking *space weather* [6]. The geomagnetic field of the Earth prevents the electrically charged particles of the solar wind to enter the surroundings of the Earth.

Solar wind interaction with the magnetosphere is reflected by the fact that by the action of the Lorentz force are electrically charged particles moving in the direction of the magnetic field lines and flow around the magnetosphere. The solar wind thus distorts the shape of the magnetosphere (Figure 4): the front (daily, windward) side of the magnetosphere is compressed to about ten Earth radii (about 60.000 km) and dark (night, leeward) side is extended into a long tail of magnetosphere which extends far beyond the orbit of the moon. Geomagnetism thus shields Earth from the solar wind and thus protects the Earth's biosphere. The existing forms of life would not exist on our planet without geomagnetism. Just around the geomagnetic poles, where the magnetic induction vector is approximately perpendicular to the surface of the Earth, channels through which the solar wind in particles can partially penetrate into the deeper layers of the magnetosphere and endanger with radiation passengers and crew of transoceanic flights are created. A statistically significant incidence of skin cancer among pilots was proved.

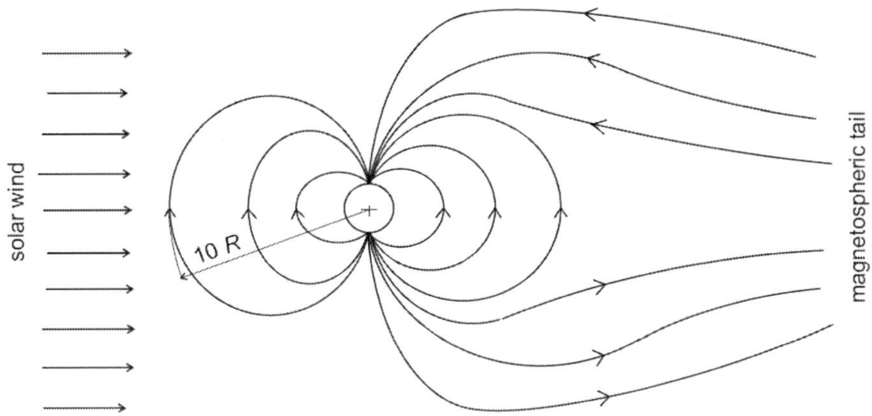

Figure 4. Magnetosphere deformed by the solar wind strain.

If CEM, that hits the Earth's magnetosphere with the solar wind, has a sufficiently high energy, the geomagnetic activity is reflected not only in the polar regions, but some of this energy is transferred to the magnetosphere, resulting in rapid variations of the geomagnetic field. Their intensity is different: from minor defects, which occur several times a day, to strong variations, caused by the solar wind in shock wave – called *magnetic storms*. There are usually accompanied by polar lights. Minor daily variations in the time of quiet geomagnetic field are from 20 to 30 nT and during the day, resp. year show some periodicity. For big magnetic storms they reach up to hundreds nT. The size of the geomagnetic field is classified on a ten scale and is referred to as K0 to K9. For example, when K0 is the magnetic flux density $B \in \langle 0, 5 \rangle$ nT, at K3 is $B \in \langle 20, 40 \rangle$ nT a the K9 represents $B > 500$ nT.

Solar eruptions are one of the manifestations of physical processes that take place not only on the Sun, but also on other stars, which were identified much more powerful eruption than the sun. Were called *supereruption* and their energy can be up to four orders of magnitude more massive than the Sun, i.e., reach values 10^{27}– 10^{28} J. Astrophysicists are investigating whether we can expect the emergence of these supereruption also on Sun. Systematic study and forecasts of these phenomena is engaged in global research *Space Weather Prog*Ram. Statistical analysis of the results of observations of eruptions on the plurality of stars that have the size and period of rotation similar to the Sun (i.e., a star-type G) shows that the frequency of occurrence of eruptions decreases with an increase of power by the power law [7, 8, 9]. Based on the processing of data from NASA's Kepler occurs on our Sun, according to [9], the eruption of energy 10^{25} J once a year, while supereruption of energy in 19^{28} J, which would endanger life on Earth, have a frequency from 800 to 5000 years. Any contact with the Earth's magnetosphere by the significant object of CME with consequent development of geomagnetic storms is beyond our control, but we try to construct mathematical models to simulate this natural phenomenon and to predict it to give enough time to perform technical measures to prevent greater damage.

3. Definition of the Problem

This work will show how to calculate parasitic currents that can be induced into electricity network during geomagnetic storms. The question of the influence of variations of the geomagnetic field on the electricity supply system, caused by solar activity, became current at the end of the last century, in connection with several big accidents. Examples include massive power supply black-out on the east coast of North America in October 1989. The largest observed solar flare followed by a strong magnetic storm ($B \sim 3000$ μT) registered British astronomer R. Carrington in August 1859. Since the introduction of electricity into practice at that time was at the very beginning, although magnetic storm knocked out the telegraph network in the entire northern hemisphere, it did not cause any more damage. Another recent example is a massive solar flare (energy ~ 10^{25} J), which caused a magnetic storm, during which there was a collapse of electrical networks in areas of Quebec in Canada. At the present time, when the entire human civilization depends on electrical network and electronics, similar storm might cause incalculable damage. List of magnetic storms, which damaged electrical equipment in different parts of the Earth, is given in [10].

Research into these phenomena is discussed by many authors, among whom a significant proportion belongs to Finnish authors [11]. In addition to electrical systems, telecommunications networks, network for transport oil and gas and the rail track were also considered. Despite great efforts that have been devoted to resolving this issue, a reliable prediction of the effects of magnetic storms is not yet closed.

Parasitic currents induced by quickly geomagnetic variations into the branches of the transmission network – called in abbreviation the Geomagnetically Induced Currents (GIC) – can be harmful to electricity networks. Geomagnetic variations have the character of shock waves, which change in seconds. Geomagnetic variations have the character of shock waves, which change in seconds. GIC are superimposed together with the operating currents in the lines of electricity system transmitting active and reactive power. In terms of electrical networks, these changes are very slow and therefore GIC are denoted as quasi-DC currents. It follows that it is sufficient to count only with the resistance of the transmission lines, while the effect of inductance, capacity coupling and skin effect are negligible. GIC cause increase the load of conductors by current, increase the transmission losses, increase the voltage drop in the lines and cause oversaturation of magnetic circuits of transformers and thus, they may cause a violation of stability and eventually black-out of system [12].

4. Algorithm for Calculating the Effect of Geomagnetic Storms on the Electrification System

4.1. Geomagnetic Field on the Surface of the Earth and Its Rapid Variations

The rate of time of vertical components of $B_n(t)$ of vector of magnetic flux density $\boldsymbol{B}$ is important for electromagnetic induction GIC in the high-voltage power distribution. The magnetic storm on Earth is delayed from the eruption on the Sun for about 25 – 60 hours. Based on many years of observations of solar and geomagnetic variations, the results of subsequent measurements can $B_n(t)$ be predicted with a good probability. Estimated time evolution of $B_n(t)$ will be the input parameter for the calculation algorithm of GIC, which will be described in the next section.

4.2. Calculation of GIC in a Passive Network

We will not consider transformers and wires in the high-voltage distribution network and restrict ourselves to conductors of one phase. We obtain passive network $\mathcal{N}$, distributed in the plane (x, y), and composed of n irregular polygons and containing m branches and p nodes. A time-varying geomagnetic field acts on the network $\mathcal{N}$; we know its vertical component B_n, which has the character of the pulse length T and we will assume that it does not change in the whole network $\mathcal{N}$. If the network $\mathcal{N}$ is so large that this assumption is not satisfied, the algorithm described below can be easily generalized. Methods of matrix analysis of electrical circuits, which are described e.g., in [13] and [14], will be used.

Topological structure of the polygonal network $\mathcal{N}$, we express by the planar graph $\mathcal{G}$ (Figure 5), which then algebraically displayed using the incidence matrix A[n, m]. Into all n polygons graph $\mathcal{G}$ we introduce loops $\{c_1,\ldots, c_n\}$. Branches and loops of the graph $\mathcal{G}$ can be arbitrarily oriented. It can be proved (see eg. [13, 14]), the equation we received by applications of Kirchhoff's voltage law on these n loops form a system of linearly independent equations. Therefore, loops $\{c_1,\ldots,c_n\}$ is called a *system of independent loops*. Incidence matrix A[a_{ij}] is rectangular ($i = 1,\ldots, n$; $j = 1,\ldots, m$) and expresses incidence between branches oriented graph $\mathcal{G}$ and its oriented loops. The rows correspond to the loops and the columns to the branches of the graph $\mathcal{G}$. The elements of matrix **A** are: $a_{ij} = +1$, if the i-th loop is positively incident with j-th loop, $a_{ij} = -1$, if the i-th loop is oppositely incident with j-th loop and $a_{ij} = 0$, if the i-th loop is not incident with j-th loop ($i = 1,\ldots,n$, $j = 1,\ldots, m$). Since the matrix A have been prepared for a set of independent loops, thus has rank $h = n$. The assignment of matrix A to the graph $\mathcal{G}$ is unambiguous; the converse does not apply.

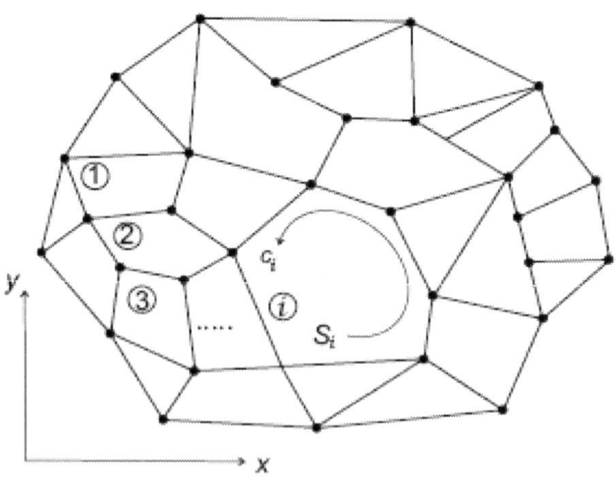

Figure 5. Graph $\mathcal{G}$ of the polygonal passive network $\mathcal{N}$.

Metrical structure of the network $\mathscr{N}$ define the coordinate $(x_1, y_1),\ldots, (x_p, y_p)$ their nodes, i.e., vertices of polygons a network. The contents of these n polygons may be computed by the known formulas. Let i-th polynomial has vertices $(x_1, y_1),\ldots$. Then its contents S_i is

$$S_i = \frac{1}{2}\left|\left\{x_1(y_2 - y_p) + x_2(y_3 - y_1) + x_3(y_4 - y_2) + \ldots + x_p(y_1 - y_{p-1})\right\}\right| = \\ = \frac{1}{2}\left|\left\{y_1(x_2 - x_{p-1}) + y_2(x_3 - x_1) + y_3(x_4 - x_2) + \ldots + y_p(x_1 - x_{p-1})\right\}\right| \tag{1}$$

Physical structure of the network $\mathscr{N}$ is expressed as its branches-resistances R_i $(i = 1,\ldots, m)$.

According to the second Maxwell's equations applied to the i-th loop c_i

$$\oint_{c_i} \boldsymbol{E}\,\mathrm{d}\boldsymbol{l} = -\frac{\mathrm{d}\Phi_i}{\mathrm{d}t}, \quad i = 1,\ldots, n \tag{2}$$

where $\boldsymbol{E}$ is the electric field intensity and the magnetic flux linked with the loop c_i is

$$\Phi_i = \int_{S_i} B_\mathrm{n}\,\mathrm{d}S = S_i B_\mathrm{n} \tag{3}$$

Equation (2) can be written in our case in the form

$$\sum_{k} \pm u_{ki} = S_i \frac{\mathrm{d}B_{ni}}{\mathrm{d}t}, \quad i = 1,\ldots, n \tag{4}$$

where u_{ki} is the voltage of the k-th branch linked with i-th loop c_i. Equation (4) can be written in the matrix form:

$$\mathbf{Au} = \mathbf{F}(t) \tag{5}$$

Here, $\mathbf{A}$ (n, m) is incident matrix expressing the topology of the oriented graph $\mathscr{G}$, $\mathbf{u}$ is the matrix of the branch voltages

$$\mathbf{u} = \begin{bmatrix} u_1 \\ \dots\dots \\ u_m \end{bmatrix} \tag{6}$$

and $\mathbf{F}(t)$ is the matrix of loop induced voltages of the network $\mathcal{N}$

$$\mathbf{F}(t) = \mathbf{S}\,\frac{\mathrm{d}B_\mathrm{n}}{\mathrm{d}t}, \tag{7}$$

where **S** is the matrix of content of polynomials

$$\mathbf{S} = \begin{bmatrix} S_1 \\ \dots\dots \\ S_n \end{bmatrix}. \tag{8}$$

We implement the matrix of the branch currents **i** and the matrix of the loop currents **i´** in n independent loops:

$$\mathbf{i}(t) = \begin{bmatrix} i_1 \\ \dots\dots \\ i_m \end{bmatrix}, \quad \mathbf{i}'(t) = \begin{bmatrix} i'_1 \\ \dots\dots \\ i'_n \end{bmatrix} \tag{9}$$

Both matrices **i** and **i´** are related through by the incident matrix **A** (for more details see e.g., [13, 14]:

$$\mathbf{i} = {}^{\mathrm{t}}\mathbf{A}\,\mathbf{i}' \tag{10}$$

where the superscript „t“ indicates the transposed matrix.

The vectors **u** and **i** are related through the Ohm law in the matrix form:

$$\mathbf{u} = \mathbf{R}\,\mathbf{i} \tag{11}$$

where **R** is the diagonal matrix of the branch resistances:

$$\mathbf{R} = \mathrm{diag}\,[R_1,\ldots, R_m] \tag{12}$$

Equation (11) will be multiplied from the left with the matrix **A** and the vector **i** will be substituted according to the equation (10):

$$\mathbf{Au} = \mathbf{ARi} = \mathbf{AR}\,{}^{\mathrm{t}}\mathbf{Ai}' \tag{13}$$

Using the equations (5) and (13) we obtain

$$\mathbf{F}(t) = \mathbf{AR}\,{}^{\mathrm{t}}\mathbf{Ai}' \tag{14}$$

This equation serves for computation of the vector $\mathbf{i}'$ of loop currents and using the equations (10) and (7) we obtain the matrix of instant values of GIC in branches of the network $\mathcal{N}$ as

$$\mathbf{i}(t) = {}^{\mathrm{t}}\mathbf{A}\left[\mathbf{AR}\,{}^{\mathrm{t}}\mathbf{A}\right]\mathbf{S}\frac{\mathrm{d}B_{\mathrm{n}}}{\mathrm{d}t} \tag{15}$$

From the matrix $\mathbf{i}(t)$ it is easy to determine the matrix of RMS values of the currents in branches:

$$\mathbf{I}_{\mathrm{RMS}} = \sqrt{\frac{1}{T}\mathbf{i}(t)^2\,\mathrm{d}t} \tag{16}$$

where T is the duration of the magnetic impulse.

4.3. Illustrative Example

The network $\mathcal{N}$ lying in the plane (x, y) has the topological structure characterized by the oriented graph $\mathcal{G}$ according to the Figure 6. Its metric structure is given by the coordinates of the nodes of the graph $\mathcal{G}$: 1(0, 0), 2(80, 30), 3(180,20), 4(140,120) and 5(60,160) [km]. The physical structure of the network $\mathcal{N}$ is given by the resistance of the branches: $R_1 = 20.5\ \Omega$, $R_2 = 24.5\ \Omega$, $R_3 = 41.7\ \Omega$, $R_4 = 31.9\ \Omega$, $R_5 = 26.3\ \Omega$, $R_6 = 26.3\ \Omega$ and $R_7 = 21.6\ \Omega$. The vertical component of the magnetic induction of the geomagnetic field was determined by the measurement with the time step of 1 s; the graph of the function $B_n(t)$ is in Figure 7-top. We will determine the instant and RMS values of the GIC in the network branches.

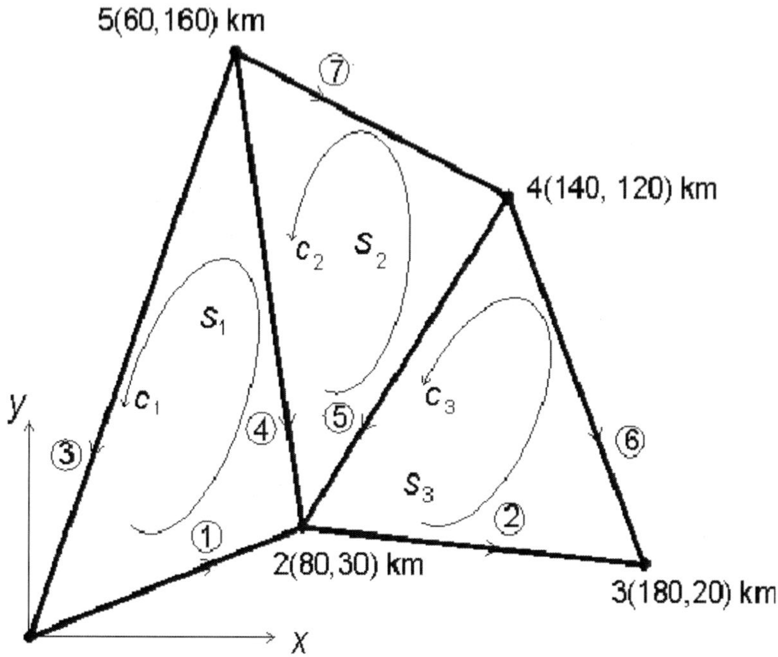

Figure 6. Passive network.

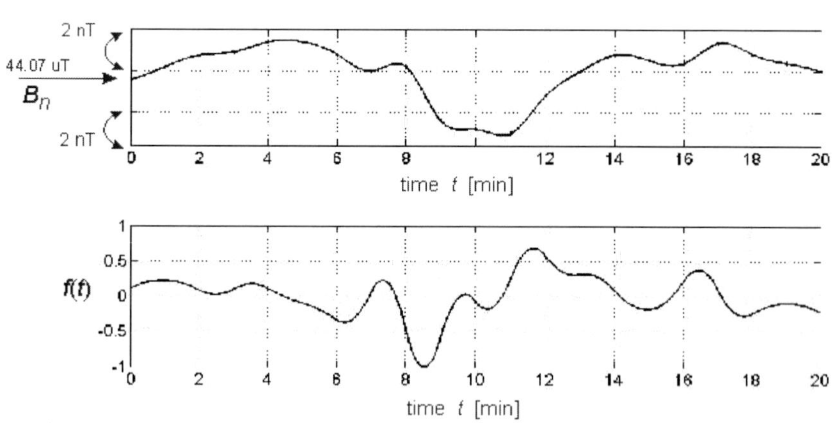

Figure 7. The time function of the normal component of the magnetic flux density $B_n(t)$ and the function $f(t)$.

Now we set instant and effective values of the currents in the branches (GIC). Using the spline interpolation followed by the numerical derivation of the function $B_n(t)$, the function $\mathrm{d}B_n/\mathrm{d}t$ was found. This function was normalized according to its maximal value, $\max[|\mathrm{d}B_n/\mathrm{d}t|] = 7{,}75$ [nT/s]. The graph of the normalized function

$$f(t) = \frac{1}{7.75}\frac{\mathrm{d}B_n}{\mathrm{d}t}\ \left[\mathrm{nT/s}\right] \quad f \in \langle -1, 1\rangle$$

can be seen in Fig.7 – bottom. In the interval $t \in \langle 0, 20\rangle$ [s] its RMS value $f_{RMS} = 0.3038$.

Introduce now oriented loops in the polygons (triangles) of the oriented graph $\mathcal{G}$. The incidence matrix, the matrix of resistances and the matrix of the content of network polygons **S** are, respectively,

$$\mathbf{R} = \mathrm{diag}\left[20.5,\quad 24.5,\quad 41.7,\quad 31.9,\quad 26.3,\quad 26.3,\quad 21.6\right]\ \left[\Omega\right],$$

$$\mathbf{A} = \begin{bmatrix} 1 & & 1 & -1 & & & \\ & & & 1 & -1 & & -1 \\ & 1 & & & 1 & -1 & \end{bmatrix},$$

$$\mathbf{S} = \begin{bmatrix} 5{,}5 \\ 3{,}9 \\ 2{,}7 \end{bmatrix} .10^9 \ \left[\mathrm{m}^2\right] \qquad (*)$$

Matrixes (*) are inserted into equation (15) so that the matrix of instant values of currents $\mathbf{i}(t)$ is obtained:

$$^{\mathrm{t}}\mathbf{i}(t) = {}^{\mathrm{t}}\left\{7.75\ {}^{\mathrm{t}}\mathbf{A}\left[\mathbf{A}\mathbf{R}\,{}^{\mathrm{t}}\mathbf{A}\right]^{-1}\mathbf{S}\right\}.10^{-9} f(t) =$$

$$= {}^{\mathrm{t}}\left[0.747;\ 0.567;\ 0.747;\ 0.121;\ -0.300;\ -0.567;\ -0.868\right].f(t)\ \left[\mathrm{A}\right]$$

Effective values of currents in the branches of network $\mathcal{N}$ are obtained using the value f_{RMS}:

$$^{t}\mathbf{I}_{RMS} = {}^{t}\left[\,\mathbf{i}(t)\,\right]_{RMS}$$

$$= {}^{t}\left[0.227;\quad 0.172;\quad 0.144;\quad 0.0368;\quad 0.09114;\quad 0.172;\quad 0.264\right]\quad [\mathrm{A}]$$

For an approximate control, the induced current in a square turn 50 x 50 km made of a copper wire of diameter of 10 mm (R = 44.6 Ω, S = 2500 km^2) is determined. A magnetic flux density B_n acts on the turn according to Fig. 7-top. The resultant current is $i = \dfrac{S}{R}\dfrac{\mathrm{d}Bn}{\mathrm{d}t} = 0.434\, f(t)$ and its RMS value is $I_{RMS} = 0.434\ f_{RMS} = 0.1318$ A, which is in accordance with the mentioned example.

5. Calculation of GIC in Active Network

The passive network $\mathcal{N}$ can be used as a model for calculating the GIC in the pipelines for oil or gas transportation, or for a railway of rail network. The calculation of GIC in the high-voltage electrical network is performed using active networks. It is obtained in such a manner that high power transformer windings are put into the model of a passive network and the three phases and grounding wire are considered afterwards. For more details about this dealt with in [15].

Conclusion

The time function of the magnetic flux density $B_n(t)$ is the input quantity for the calculation of GIC. This quantity can be estimated based on larger number of measurements. Thus this is the typical uncertain input data task. Task like these can be generally solved in two different ways: stochastic (probability task), or by the so called reliable solution method (thus the method of the worst case), when we proceed from the "worst" possible time function of $B_n(t)$, even when the probability of its occurrence is small. The second possible approach is more pessimistic, but it is "safer".

Size of GIC is independent of the size of the normal component of the magnetic flux density B_n, but on the size of its derivative dB_n/dt. For this reason we recommended to evaluate the activity of the geomagnetic field not only by the size of the magnetic flux density on a scale K0 to K10, but also by the values of the derivative of the magnetic flux density with rerspect to time, at the similar scale.

The magnitude of the normal component of the magnetic field B_n is not decisive for the GIC; decisive is its time change, thus the derivation dB_n/dt. Based on the monitoring of the geomagnetic field during last years published in [1], it can be supposed that strong shock geomagnetic waves that may lead to the prediction of the black-outs of the electrification network, are rare events. If these extraordinary storms occur in the future, the energy network operator can prevent the destructive situations only by the intervention into the topological structure of the network. He can for example divide the distribution network into several smaller groups, or he can interrupt the galvanic connection of some parts of the network exposed to the activity by connecting the breaking transformers.

REFERENCES

[1] http://bdv.ig.cas.cz/index.html, http://rwcprague.ufa.cas.cz

[2] Glatsmaier, G. A. & Roberts, P. H. (1995). *Phys. Earth Planet. Inter.*, Vol. *91*, pp. 63.

[3] Parker, E. N. (1979). Cosmical magnetic fields. Clarendon Press, Oxford.

[4] Parker, E. N. (1963). Interplanetary dynamical processes. Wiley-Interscience, New York.

[5] Ertepinar, P., et al. (2012). Archeomagnetic study of five mounds fromUpper Mesopotamia Between 2500 and 700 BCE. *Earth and Plane. Sci. Letters.* Vol. *357-358*, Dec., 84-98.

[6] Priest, E. R. (1984). *Solar magnetohydrodynamics.* D. Reidel Publ. Co., Dordrecht (Holland).

[7] Priest, E. R. (1981). Solar flare magnetohydrodynamics. Gordon and Breach Science Publishers, New York.

[8] Shibata, K. & Magara, K. (2011). Solar flares: Magnetohydrodynamics processes. *Living Rev. Sol. Phys.*, *8* (6), 1 – 99.

[9] Schaefer, B. E. (1998). *Flashes from normal stars.Astrophys. J.*, Vol. *337*, 927-933.

[10] Boteler, D. H., et al. (1998). The effects of geomagnetic disturbances on electrical systems at the earth's surface. *Adv. Space Res.*, Vol. *22*, No.1, 17-27.

[11] Elovaara, J. P., et al. Geometrically induced currents in the nordic power systems. Paper 36-301 of the *Proceedings of the CIGRE 1992 Sesion*, 30 Aug.-5 Sept., Paris, pp. 10.

[12] Bolduc, L. & Aubin, J. (1978). Effects of direct currents in power transformers. *Electronics Power Syst. Res.*, *1*, 291-298.

[13] Mayer, D. (1966). *Analysis of electrical circuit with matrix calculus.* Academia, Prague, (In Czech)

[14] Mayer, D. (1981). *Introduction to the theory of electric circuits*. SNTL/ ALFA, Prague, 2nd ed. (In Czech)

[15] Mayer, D. (2013). A contribution to the investigation of influence of geomagnetic storms on electrification systems. *Acta Technica AV CR*, Vol. *58*, No. 4, 351 – 365.

In: Cyclonic and Geomagnetic Storms
Editor: Victoria P. Banks
ISBN: 978-1-63482-360-9

Chapter 2

RADIATION ASPECTS OF GEOMAGNETIC STORM IMPACT BELOW THE RADIATION BELT

A. V. Suvorova ***and*** *A. V.* Dmitriev
National Central University, Taiwan
Lomonosov Moscow State University, Russia

ABSTRACT

The spectrum of phenomena produced by geomagnetic storms in the near Earth's environment is gradually widened as observational data increase. One of them is significant enhancements of energetic electrons (>10 keV) revealed recently in the forbidden zone below the radiation belt. A comprehensive analysis of the forbidden electron enhancements during major and recurrent magnetic storms in solar cycles 23 and 24 was conducted on the base of experimental data acquired from low-orbiting satellites from 1998 to 2014. A possible mechanism of the phenomenon is discussed. Knowledge of where and when such environmental events will occur can be used for both space physics and applied researches. We focus on the following radiation aspects of the electron enhancement problem with a short review for each item: 1) ionization impact of 10-30 keV electrons in the ionosphere; 2) X-ray and gamma-ray contaminations by 100-300 keV electrons to astrophysical observation on board low-orbiting satellites; 3) possible effects of energetic electrons in satellite anomalies attributed to low-orbit radiation environment.

1. INTRODUCTION

The inner radiation belt (IRB) of intense fluxes of trapped energetic particles is stably located above the ionosphere (~1000 km) except the vicinity of South Atlantic Anomaly (SAA), where the lower edge of IRB reaches the ionospheric *F* region at altitudes of 300 - 600 km (see also a review by Lazutin and Kudela [2012] about the Earth's magnetosphere). The well-established general concept of the near-Earth radiation is schematically illustrates in Figure 1. Figure 1 shows mutual alignment of the IRB and ionospheric top boundary surrounding the titled and offset Earth's magnetic dipole. Outer radiation belt (ORB) reaches the ionospheric heights only at middle and high latitudes (not shown). At equator-to-low latitudes and outside SAA, the low-altitude radiation is negligible because energetic particles are effectively scattered and loss energy in interaction with atmospheric neutrals and so, this region is called a forbidden zone (see Lemaire [2003] for a review). This important circumstance is always taken into account in satellite-born astronomy/astrophysics experiments conducting at low Earth orbit (LEO) with an altitude extending the Earth's surface up to 2000 km. A low inclination orbit is preferred because the equatorial region away from the SAA is characterized by low radiation background.

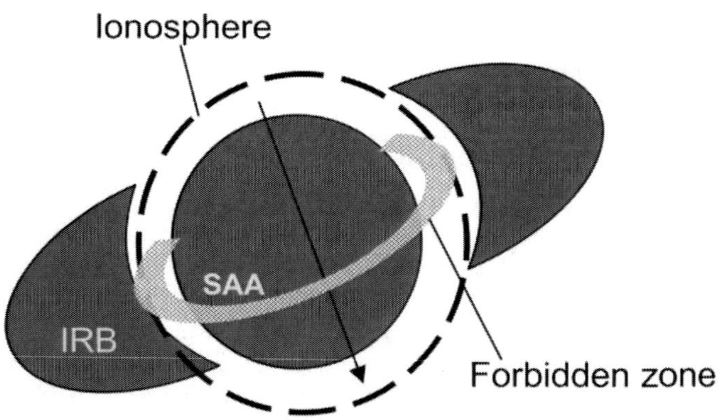

Figure 1. A sketch of the near-Earth radiation in the geodipole magnetic field: the inner radiation belt (IRB) indicated by the red color and South Atlantic Anomaly (SAA). Forbidden zone (gray ring) is placed below IRB in the ionosphere.

Since the 2000th when LEO multi-satellite measurements of radiation in the near-Earth environment have begun, it became possible to reveal unexpected behavior of the energetic electrons below the IRB and carry out statistical studies of this. During strong magnetic storms, energetic electron fluxes suddenly increase by 4-6 orders of magnitude at altitudes 600 - 900 km in the forbidden zone, then drift for several hours in east direction to the SAA and finally are lost there [Asikainen and Mursula, 2005; Dmitriev and Yeh, 2008; Suvorova et al., 2012a]. Measurements have showed that pitch angles of energetic electrons range within the drift loss cone, i.e., electrons are locally trapped or in other words the "quasi-trapped". Fluxes of precipitating electrons in the forbidden zone remain stably low in most cases. Sometimes they weakly increase, by only 2 orders of magnitude, but then rapidly decay within 1-2 hours due to atmospheric scattering [Suvorova et al., 2013a]. Because the quasi-trapped electron fluxes enhanced in the forbidden zone we call them forbidden energetic electrons (FEE).

Actually, the phenomenon of energetic particle enhancements below the IRB was known from 1960s and the problem was discussed in literature (see Suvorova et al. [2013a] and Suvorova and Dmitriev [2014] for references). However, sparse observations of sporadic events by a single satellite caused strong argument due to a doubt about validity of measured high intensity (see review by Paulikas [1975]). As a result of this, further investigation of the FEE phenomenon was not carried out.

For better understanding a spatial and temporal behavior of FEE, let's consider particle motion along a drift L shell and locations of its mirror points at different longitudes. Figure 2 shows the longitudinal variation of the height of the drift L shells calculated for the geomagnetic equator using the IGRF model for the epoch of 2005. Due to the asymmetrical orientation of the geomagnetic dipole, the height of the top points of the magnetic field line varies with longitude, and heights of mirror points in the northern and southern hemispheres also differ. Locally trapped electron with a pitch angle of ~90° has mirror points near a top point of the magnetic field line, so it bounces along magnetic field line about this point. The bouncing motion is very fast and has a period of a portion of a second [Lyons and Williams, 1984].

Nonuniformity of the geomagnetic field accounts for slower longitudinal drift of the electrons to the east for time ranging from minutes to days. The drift L shells reach minimal heights in the SAA region, where practically all of the particles, trapped at $L < 1.1$, are lost due to atmospheric scattering. Thus, quasi-trapped electrons or FEE do not complete the circle of azimuthal drift around the Earth because they descent in the SAA and are eventually lost in the atmosphere. Therefore, it is reasonably to expect that duration of an enhanced FEE-flux phenomenon should be less than the azimuthal drift period of electrons of a given energy. Actually, in many cases observations show much longer durations of FEE enhancements that contradict the limited time of drifting [Suvorova et al., 2014b].

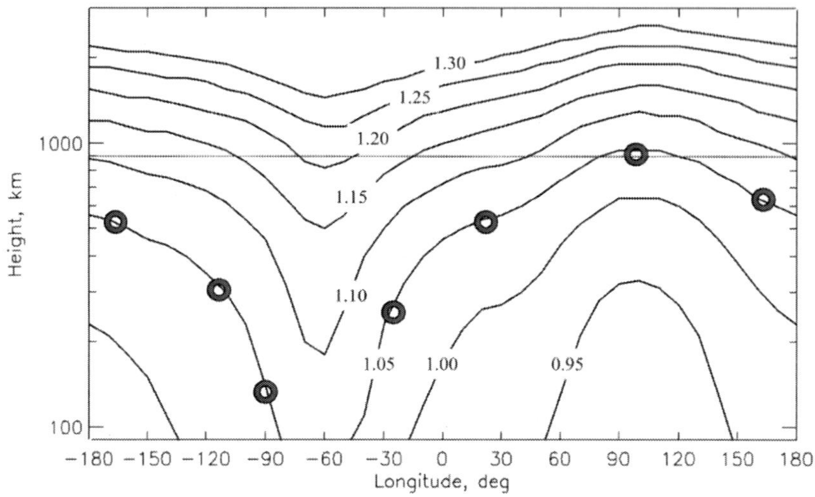

Figure 2. Drift shells (*L*-shells) of energetic particles in the coordinates of height vs longitude. *L*-shells were calculated at geomagnetic equator from IGRF model for epoch of 2005. Horizontal dashed lines indicate the 900 km height, which is close to an orbit height of the NOAA satellites. Red symbols indicate $L \sim 1.05$, approximately an upper edge of the forbidden zone. Forbidden (or quasi-trapped) energetic electrons drift eastward and are lost in the SAA region located in the range of longitudes from –100° to 0°.

We have studied features of spatial and temporal behavior of FEE using large statistics of multi-satellite measurements [Suvorova et al., 2012b, 2013a, b, c, 2014a, b]. Main results of the work are presented here. In this work, we focus on the following important issues concerning the nature and physics of the phenomenon, and its impact on the ionosphere: (1) relation FEE enhancements to geomagnetic activity (storms/substorms), (2) a source and mechanism of FEE enhancements, (3) ionizing particle effects in the atmosphere-ionosphere system. We touch briefly application issues and radiation aspects of FEE affecting satellite systems that is contributions to astrophysical satellite-based measurements, instrumental effects, and affecting scientific equipment and electronics on board spacecrafts in LEO.

Table 1. Numbers of the POES satellites operating during a year and FEE-associated days

1998	1999	2000	2001	2002	2003	2004	2005	2006	2007	2008	2009	2010	2011	2012	2013	2014
1	1	1	2	3	3	3	4	4	5	5	6	6	6	6	6	4
2	6	8	8	5	>50	>50	>100	~100	~40	~60	>40	~50	~60	~40	~20	~10

2. Enhancements of Forbidden Energetic Electrons (FEE)

2.1. Observations of Energetic Electrons by a Fleet of the NOAA/POES Satellites

Energetic electrons at low altitudes are continuously monitored from 1990s by a fleet of polar low-orbiting NOAA satellites (POES) at altitudes of ~ 850 km [Evans and Greer, 2004]. We investigate dynamics of the quasi-trapped radiation in the forbidden zone during geomagnetic disturbances, storms and substorms, using the huge database at http://satdat.ngdc.noaa.gov.

The POES satellites supply data on the trapped and precipitating electrons in the low/medium energy range from 30 keV to 300 keV for almost 2 solar cycles [Rodger et al., 2010]. The POES satellites have Sun-synchronous orbits. Number of satellites was extended from 1 to 6 step-by-step (Table 1, second row) that allowed increasing the orbital planes at different local times. Note, that during the last year, 2014, only 4 satellites operated because two satellites had failed. For first three years (1998 – 2000), a single satellite NOAA-15 was operating and hence, temporal and spatial coverage of measurements was sparse.

At first, we show geographic distributions of the energetic electrons during quiet and storm days, which agree with the conventional picture in Figure 1. Figure 3 presents 1-day global maps of the >30 keV electron flux intensities measured by POES satellites under different geomagnetic conditions: (left) a "perfect" quiet day on 17 July 2009 without any notable geomagnetic activity, and (right) a major geomagnetic storm on 30 August 2004 with the maximal *Dst* index (or SYM-H) of about -120 nT. The maps were compiled from multi-satellite measurements by pairs of orthogonally oriented detectors: 0°-detector pointed to zenith (upper panels), and 90°-

detector pointed to north-south direction (bottom panels). In each spatial bin, the maximal value of electron flux is shown.

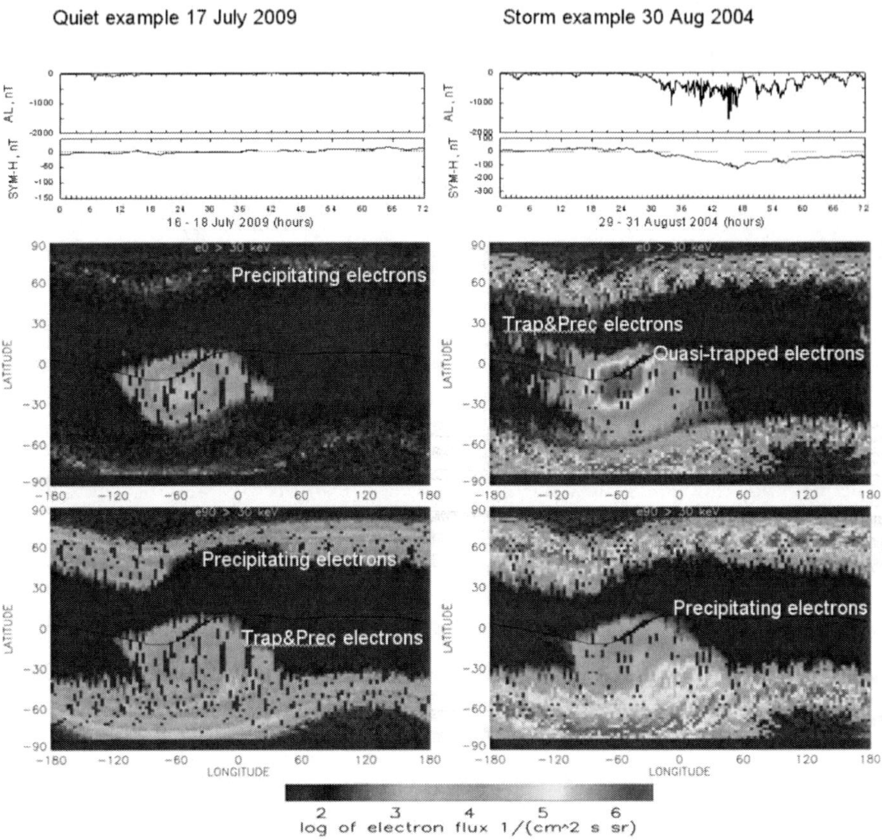

Figure 3. Geomagnetic conditions, the SYM-H (or *Dst*) and *AL* indices (upper panels), during a quiet day on 17 July 2009 (left) and a storm day on 30 August 2004 (right) and geographic maps of >30 keV electron fluxes: (upper maps) 0º-telescope and (lower maps) 90º-telescope. These maps are compiled from multiple-satellite NOAA/POES data during the quiet and storm days, respectively. Dip equator is indicated by the black curve.

At high latitudes, both detectors measured only fluxes of precipitating electrons. At low-to-mid latitudes both detectors measured a mix of trapped and precipitating populations. At the equator, the 0°-detector measured only quasi-trapped electrons (upper panels) and the 90°-detector measured only precipitating electrons (bottom panels). As one can see in the upper and bottom panels, the forbidden zone is characterized by tenuous fluxes of the quasi-trapped and precipitating electrons <100 particles per (cm^2 s sr) for the both, quiet and storm conditions. During the geomagnetic storms (right panels), the precipitating and trapped electron fluxes at high latitudes and in SAA (auroral zone and radiation belt) increase by several orders of the magnitude, from 10^4 units up to 10^7 units (where units mean electrons/(cm^2 s sr)).

2.2. FEE Enhancements during Transient and Recurrent Storms

The 17-year statistics shows that the radiation in the forbidden zone is not always steadily low. Namely, fluxes of the quasi-trapped energetic electrons within the wide energy range exhibit highly dynamical behavior. They can increase dramatically by a few orders of magnitude relative to the quiet level [Suvorova et al., 2013a]. This picture doesn't conform to a general view (Figure 1) that is a significant population of trapped electrons cannot appear within the low-latitude region not only during quiet time but also during storms.

We have found that the FEE enhancements often occurred (but not always as it is shown in Figure 3, right) during geomagnetic storms, both major and minor ones. Figure 4 shows 1-day global maps with FEE events for two types of geomagnetic storm: transient or CME-driven storm on 15-16 May 2005 (left) and recurrent or CIR/HSS-driven storm on 28 July 2006 (right). Here we use conventional acronyms for following terms: a coronal mass ejection as CME, a corotating interaction region as CIR, and a high-speed solar wind stream as HSS.

The FEE fluxes during the 15 May 2005 super-storm (*Dst* ~ -300 nT) are most intense and long persisting [Suvorova et al., 2014b]. The great FEE enhancements appear in all three energy bands >30 keV, >100 keV, and >300 keV (from top to bottom). The electrons of lower energy outside the SAA region achieved an extremely large value of $3 \cdot 10^7$ units, as much as in the IRB (including SAA) and auroral zone with ORB. The enhancements in >30 keV and >100 keV were long-lasting. The most prolonged equatorial enhancement in >30 keV occupied the forbidden zone at L shell of 1.05 – 1.15 for more than 20 hours, and >100 keV for 10 hours. Fluxes of electrons with higher energies, >300 keV, show a few burst-like enhancements of moderate intensity 10^4 – 10^6 units, and apparently due to their relatively fast azimuthal drift they first disappear in SAA.

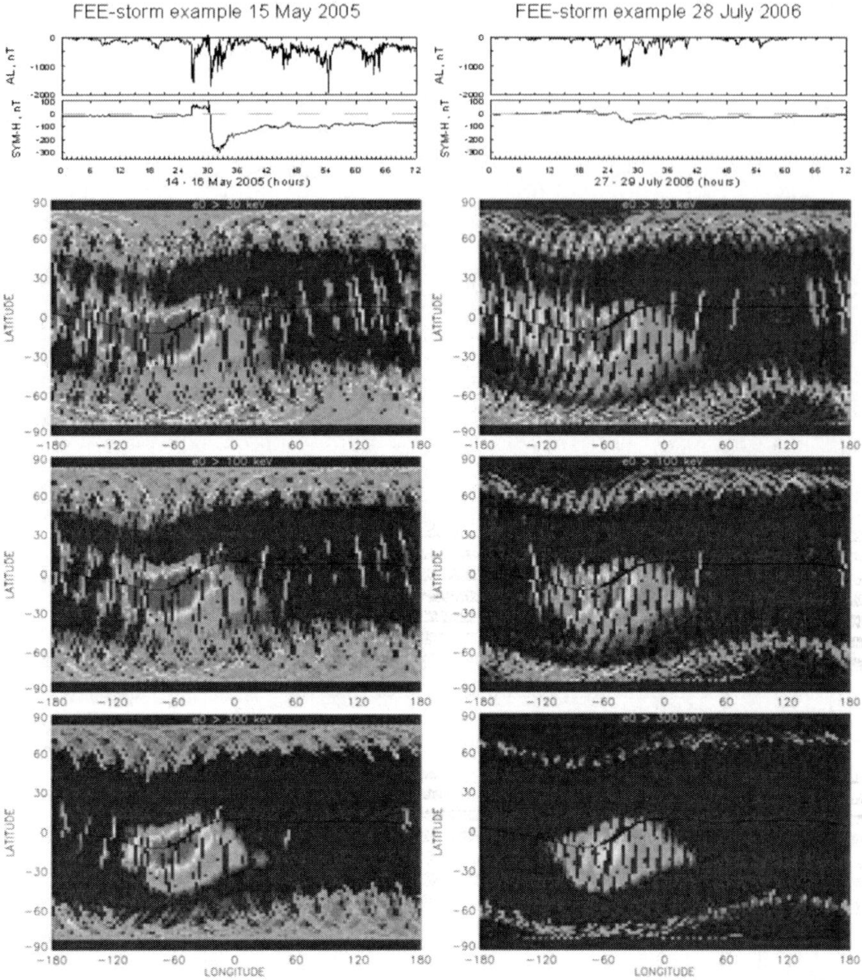

Figure 4. Geomagnetic conditions, the SYM-H and *AL* indices (upper panels), during a super-storm on 15-16 May 2005 (left) and a recurrent storm on 27-28 July 2006 (right) and geographic maps constructed from 0°-telescope data in three energy ranges: >30 keV (second panels from the top), >100 keV (third panels from the top) and >300 keV (bottom panels). Dip equator is indicated by the black curve. FEE enhancements of >30 keV and >100 keV energies are observed in both cases.

Strong FEE fluxes of < 10^6 units occurred during the 28 July 2006 minor storm with *Dst* ~ -70 nT but the enhancements were observed only in lower energy bands >30 keV and >100 keV (Figure 4, right). Comparing global maps for the 28 July 2006 minor storm (Figures 4, right) and the 30 August

2004 strong storm (Figure 3), we should note that the last one did not result in FEE enhancement at all.

Figure 5 shows 1-day global maps with FEE events during a recurrent storm on 30 August 2009 and a transient storm on 3 August 2010 occurred around the solar minimum between cycles 23 and 24. For the recurrent storm (left), one can see that the FEE enhancements are quite strong with a flux exceeding 10^5 units, even if the *Dst* maximum achieved only -40 nT. On the other hand, stronger transient storm (right) with the *Dst* maximum of ~ -80 nT results in weaker fluxes of ~10^4 unit. Hence, it is doubtful whether the storm-time *Dst* controls the FEE enhancements.

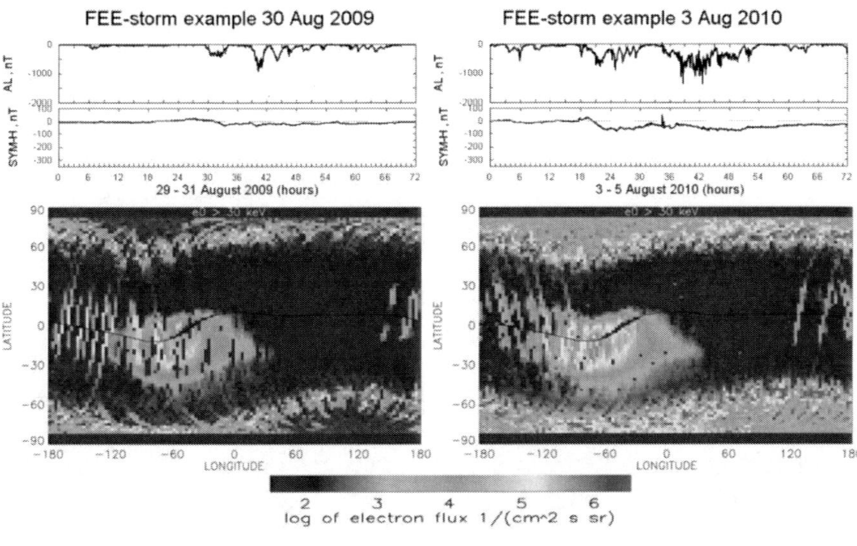

Figure 5. Geomagnetic conditions, the SYM-H and *AL* indices (upper panels), during a minor recurrent storm on 30 August 2009 (left) and a strong transient storm on 3-4 August 2010 (right) and geographic maps of corresponding FEE enhancements in >30 keV energy range. The FEE fluxes are more intense during the weaker storm.

Additionally, we make qualitative comparative analysis of a possible role of the substorm (auroral) activity in the events presented in Figures 3-5 (*AL* index, upper panels). As one can see, FEE intensities do not change directly with maximal values of the *AL* indices. There is no clear pattern of relationship between FEE intensity and the auroral activity level.

Thus, presented examples do not demonstrate unambiguous relation between enhanced FEE fluxes and the strengths of storm or substorm. We can suggest that a relation seems to be rather complicated and may be dependent on preconditioning of the magnetosphere.

2.3. Storm-Time FEE Enhancements: Pre-Storm, Main and Recovery Phases

We have found many events, when FEE enhancements occur several times during a single storm at any storm phase: initial, main and/or recovery. An example of such event is presented in Figure 6. The major transient storm on 14-15 December 2006 was previously analyzed in [Suvorova et al., 2012a, 2013a, 2015]. The geomagnetic storm started at about 14 UT on 14 December 2006, when a CME-associated interplanetary shock affected the Earth's magnetosphere. The storm initial phase was lasting until ~23 UT.

Figure 6 shows storm-time dynamics of the FEEs. Global maps contain data from 0°-detector, which measured the quasi-trapped electrons near the equator. A left panel shows electron fluxes for 3 quiet days on 2-4 December 2006. Other three maps show FEE enhancements in three storm-time intervals: the initial phase, the prolonged storm maximum and early recovery phase, and the late recovery phase. Note, that during the main phase at 23 - 01 UT, enhanced FEE was not observed. In a bottom panel of geomagnetic indices, strongly enhanced auroral activity in *AL* preceded the beginning of the first

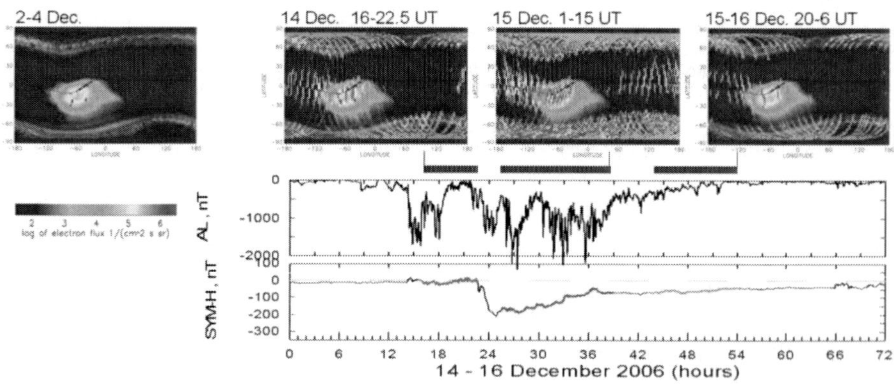

Figure 6. Storm-time FEE enhancements during different phases of a storm on 14-16 December 2006 (right panels): initial (green curve on the SYM-H panel), recovery (red curve) and late recovery (blue curve). Quite-time global map for 2-4 December 2006 is shown for comparison (left panel).

two intervals with intense FEEs, while the later period, after 15 UT on 15 December, is characterized by the gradual decreasing of the auroral activity.

Second panel shows the first FEE enhancement, which appeared between 16 and 22.5 UT during the initial phase on 14 December. This period is indicated by a green curve at the *Dst*-panel. During the storm initial-phase, several substorms (*AL* ~ -1500 nT) occurred between 14 and 18 UT. After 18 UT, the auroral activity decreased because *Bz* IMF turned northward. This pre-storm FEE event is characterized by a large enhancement up to 10^5 units and duration of 6 hours.

After 23 UT, the main phase of severe geomagnetic storm began. The storm maximum with *Dst* ~ -150 nT and Kp ~ 8+ was observed after midnight of 15 December. During the main phase at 23 - 01 UT, the FEE enhancement was not observed (not shown).

Only at the end of the main phase, the enhanced FEE fluxes again appeared within a wide longitudinal range occupying both the Eastern and Western hemispheres (third panel). The long-lasting strong FEE enhancements (~14 hours) continued during the recovery phase until 15 UT on 15 December (the red curve at the *Dst*-panel). Maximal fluxes sometimes exceeded 10^6 units.

The later period is characterized by the gradual recovery phase and decreasing of the auroral activity. Nevertheless, after 5-hour period of lack of FEE enhancements (not shown), one more enhancement was observed from 20 UT on 15 December to ~6 UT on 16 December (right panel and the blue curve at the *Dst*-panel). The last enhancement event was weaker (< 10^4 unit) and confined within a more narrow range of longitudes (only Western hemisphere), and has a clear intensity minimum near the equator. There is a split in the FEE structure into a pair of enhancements located symmetrically off the equator. Specific double-peak FEE structures are contributed by electrons in vicinity of mirror points located at the satellite height. For those electrons, the top of magnetic field line is located above the satellite orbit.

In the case presented, the FEE enhancements were observed at three consequent days with total durations of around one and half of a day. There are many other examples with a chain of FEE-events persisted during much longer period, even up to months.

2.4. Non-storm FEE Enhancements

It is interesting that FEE events can be observed also during relatively weak geomagnetic activity and without any storm. Figure 7 shows one of the non-storm FEE events occurred on 23 December 2006. During that day, as well as during previous and next days, the *Dst* value was almost constant of about -30 nT. On 23 December, the *AL* index increased to -600 and shortly peaked at -1000 nT, the *Kp* index varied within 2-4. The previous day is characterised by a larger peak value of the *AL* index of around -1500 nT.

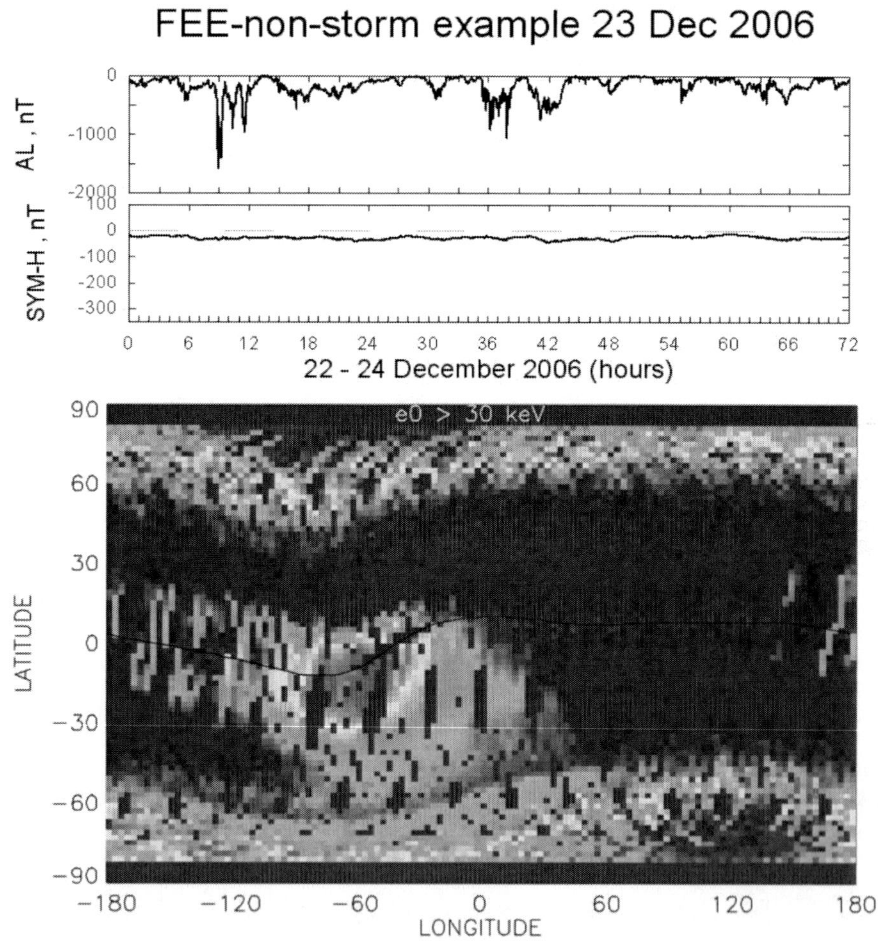

Figure 7. Non-storm FEE enhancement of >30 keV electrons on 23 December 2006. The strongest substorm in *AL* index occurred on the previous day on 22 December.

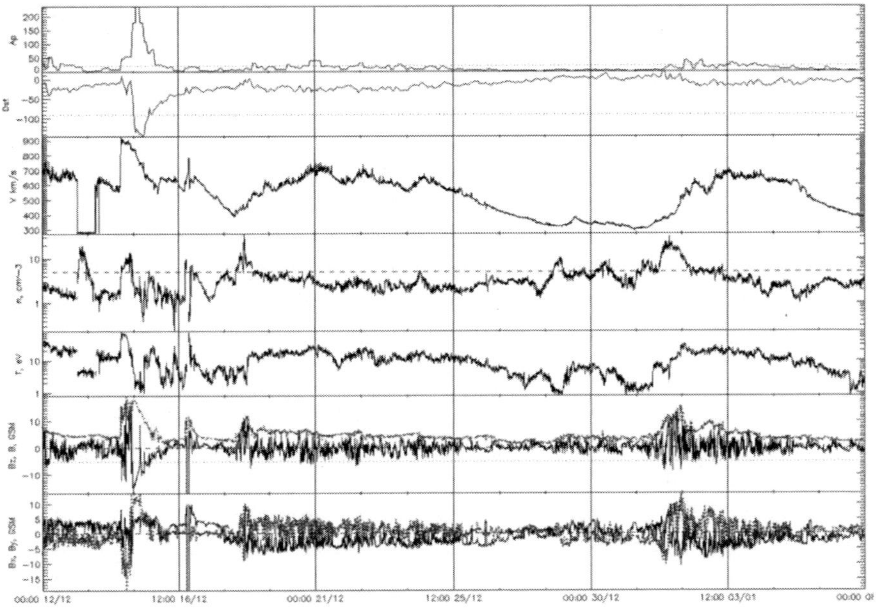

Figure 8. Geomagnetic, solar wind plasma and IMF parameters for the Bartels period of 27 days from 12 December 2006 to 8 January 2007 (from top to bottom): *Ap* and *Dst* geomagnetic indices, solar wind velocity *V*, density *n*, temperature *T*, IMF strength *B* and *Bz* component, and *B*x and *B*y components in GSM coordinates.

For better understanding of factors controlling the FEE fluxes we analyze external conditions in the solar wind. Figure 8 shows the solar wind plasma and IMF parameters for the corresponding Bartels period of 27 days. In the interval between 22 to 24 December, one can see a high speed stream structure (HSS) with a faster velocity of 700 km/s, a lower density of 3-4 cm^{-3} and highly fluctuated magnetic field components.

HSS structures are recurrent events in the solar wind. So, it is easily to understand that FEE fluxes of weak or moderate intensity should occur regularly. Moreover from a statistical study, we have found that FEE events can persists for a few days due to temporary acting of unknown mechanism.

2.5. FEE Enhancements Association with Substorms

As we have showed above, the FEE events cannot be explained by a mechanism operating exclusively during magnetic storms. Now, we redirect our attention to a relation between FEE and substorms. Suvorova et al. [2014a] have analyzed temporal characteristics of FEE in the medium-energy range >30 keV. Here we present analysis of temporal characteristics of FEE in the high-energy range >100 keV and >300 keV. We found many FEEs, which repeatedly increased and decreased within a day continuing in the same manner on ensuing days (see also Suvorova et al. [2014a]).

Figure 9 presents geographical distributions of high-energy electrons in two energy ranges >100 keV (on the left) and >300 keV (on the right) during three major geomagnetic storms on 29-31 October 2003, 7-10 November 2004 and 15-16 May 2005. In three cases presented, the largest intensity of the >100 keV electrons in the forbidden zone exceeds 10^6 units. If the quasi-trapped electron cannot close the full drift shell, it is reasonably to expect that the duration of enhanced flux phenomenon should be less than a drift period of electrons. At heights of ~900 km, 100 keV and 300 keV electrons have the drift period of ~6 and ~2 hours, respectively [Lyons and Williams, 1984]. In this connection, it is interesting to note that the great enhancements in Figure 9 were observed at longitudes, which are widely separated from each other by ~180°. That indicates to a large scale and/or long duration of the phenomenon. As for higher energy range, >300 keV, one can see a few burst-like enhancements of moderate intensity ~$10^4 - 10^6$ units.

Using ten-year data of the POES's measurements, we found that durations of the enhancements varied from an hour to ten hours and also the enhancements can be observed at any local time. We analyze time variations of >100 keV electron fluxes in the forbidden zone by considering the following types of geomagnetic disturbances: a major storm, moderate storm and strong magnetospheric compression by the solar wind plasma (interplanetary shock).

Figure 10 presents time profiles of 100 keV electron fluxes observed by POES satellites from 4 to 18 UT during the major storm on 15 May 2005 (see also Figure 9, left bottom panel). The electron flux below the IRB (black curves) at some longitudes exceeded the quiet level by ~ 4 – 6 orders of magnitude. In Figure 10, one can see that sometimes the intensities in the forbidden zone approached an extremely large value of 10^7 units, for example, after 7 UT (P5) and around 11 UT (P6, P7). The maximal intensity is the same as in the IRB (the SAA area), for example, see P5 around ~11 UT and P7 around ~14 UT, and even higher than in the outer radiation belt or auroral zone (gray curves in Figure 10).

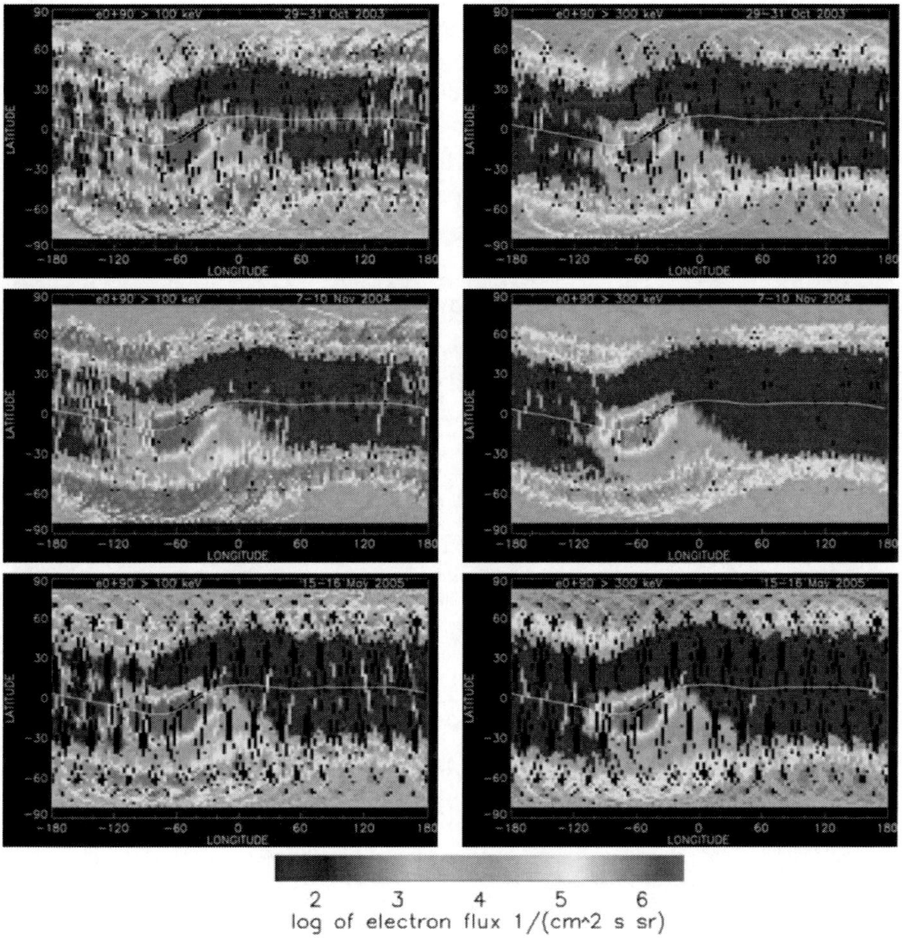

Figure 9. High-energy FEE enhancements during three super-storms (from top to bottom): on 29-31 October 2003, on 7-10 November 2004 and on 15-16 May 2005. Left panels show electrons of >100 keV and right panels - electrons of >300 keV. The largest intensity of the >100 keV FEE exceeds 10^6 units.

The duration of >100 keV electron enhancements is one of the most prolonged (about ten hours) beginning from ~7 UT and ending at ~16 UT. In Figure 10, geomagnetic activity is presented by a storm ring current (*Dst*-variation) and auroral electrojet index (*AE*), which both change in different manner. The *Dst* value varies smoothly, while the *AE*-index shows successive intensifications corresponding to substorm activations. Initially, the electron enhancements appeared during the maximum of geomagnetic activity revealed in both strong negative *Dst*-variation (~-300 nT) and very high *AE* index (~1800 nT). The geomagnetic activity, both in *Dst* and *AE* indices, was decreasing after 10 UT. But the electron enhancements were observed continuously during the next several hours on the storm recovery phase.

Another important characteristic, which we want to point out, is that the enhancements of high-energy electrons can be observed at any local time (LT), including both nighttime and sunlit side of the near-Earth orbit. Notably, the P5 satellite observed events in the early morning (6 LT) and evening (18 LT), respectively, at ~7 UT, 14 UT, 16 UT and 8 UT, 10 UT; the P6 satellite – at night (3 LT) at ~11 UT and ~1230 UT; the P7 satellite – at daytime (10 LT) and night (22 LT), respectively, at ~7 UT, 0830 UT and 0930 UT, 11 UT.

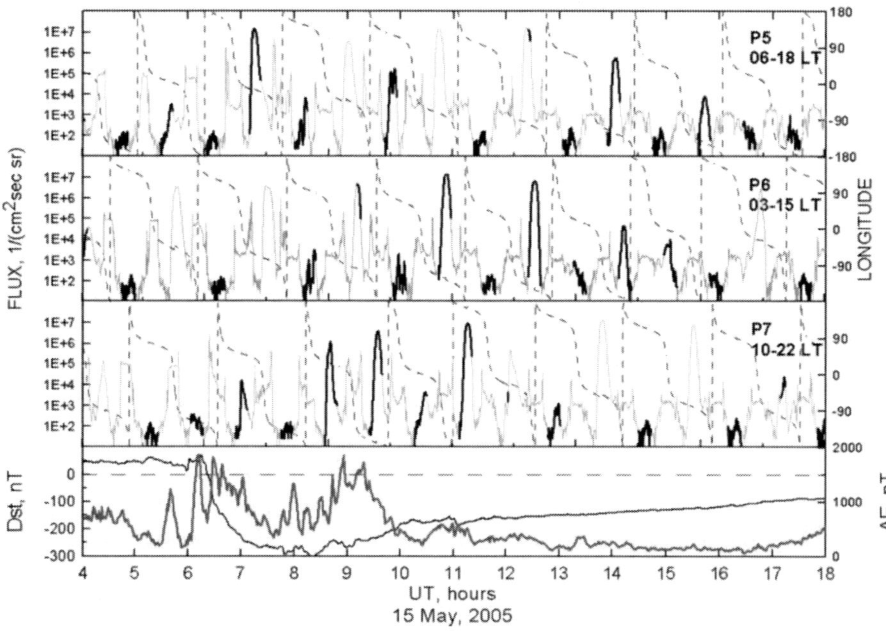

Figure 10. Time profiles of >100 keV electron fluxes observed on 15 May 2005 (from top to bottom) by NOAA-15 (P5), NOAA-16 (P6) and NOAA-17 (P7). The satellite local time is indicated on the right corners. Dashed curves indicate geographic longitudes along the satellite orbit (right axes). Black curves correspond to orbital passes at low latitudes (± 25°) and at all longitudes except the SAA area (approximately between -100° and -20°). Several electron enhancements at low latitudes are seen from 7 UT to 16 UT. The bottom panel shows geomagnetic activity: the SYM-H index (black curve, left axis) and the auroral electrojet *AE* index (red curve, right axis).

As for higher energy electrons of >300 keV, the flux increased at nighttime (2 LT and 22 LT) around 11 UT and at daytime (10 LT) around 7 UT as well (not shown). The duration of >300 keV electron enhancements was shorter than that of >100 keV electrons (about of 4 hours).

In Figure 11, we present a moderate storm event on 7-8 January 2005. We distinguish two separated time intervals, during which the >100 keV electron enhancements were observed: from 16 to 18 UT on 7 January and from 0 UT to 5 UT on 8 January (24 UT and 29 UT, respectively, in Figure 11). Particle intensities during the first interval were more than one order smaller than those during the second one. The first enhancements (see P5 and P7) were observed at local morning and prenoon (6 LT and 10 LT). Enhancements during the second interval (see P5 and P6) were observed again at local morning and afternoon (6 LT and 15 LT). Recurring enhancements at local morning can be considered as abundant evidence that both events occurred independently, because, as mentioned above, the drift period of electrons is relatively short. Note that in this event, >300 keV electrons increased only once at dusk (18 LT) closely to a vicinity of SAA (not shown). Note that a substorm (*AE* increase) occurred before each electron enhancement event.

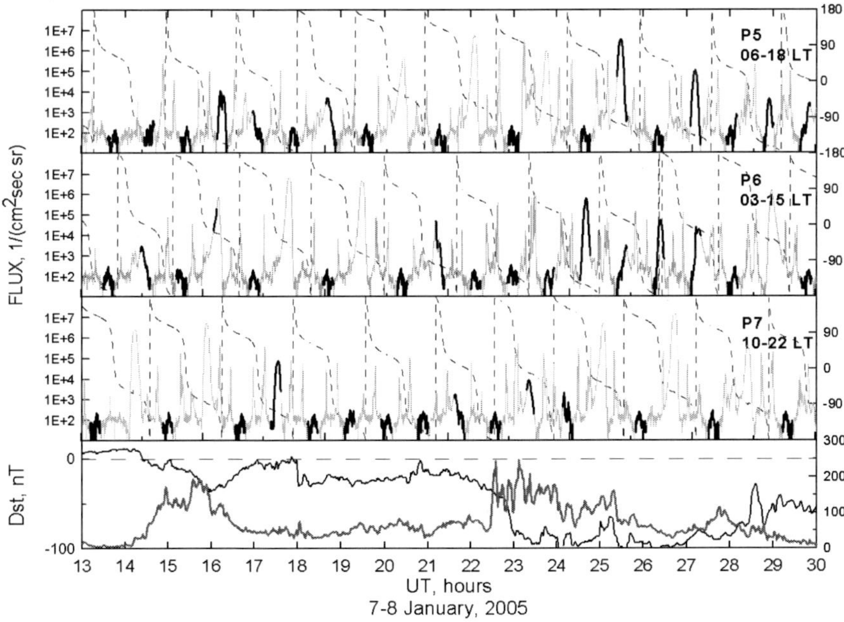

Figure 11. The same as in Figure 10 but for the moderate storm on 7-8 January 2005.

Figure 12 shows >100 keV electron enhancements during the prolonged compression of the magnetosphere by extremely high solar wind dynamic pressure of more than 150 nPa [Dmitriev et al., 2014]. Note that the magnetic storm was of moderate strength. Due to the compression, the Earth's magnetopause shrunk to about ~3-4 Re in the subsolar region, radiation belt and ring current moved closely to the Earth. The enhancements were observed at ~19 UT by P5, at 2030 UT by P7 and at 22 UT by P6, with all three occurring during daytime (6 LT, 10 LT and 15 LT). Thus, this case is a good example of single event with a short duration of about 3 hours, when the intense fluxes of >100 keV electrons disappeared within approximately three hours due to their relatively fast azimuthal drift. In this event, >300 keV electrons were not observed at all. Note also, that there was only one (isolated) substorm (*AE*-index), which is probably related to this short event.

We have demonstrated several examples of unusual latitudinal distributions of high-energy electrons during geomagnetic disturbances such as strong magnetic storms and magnetospheric compression. Figure 9 demonstrates that electron fluxes at equator-to-low latitudes extremely increase and sometimes extended almost globally.

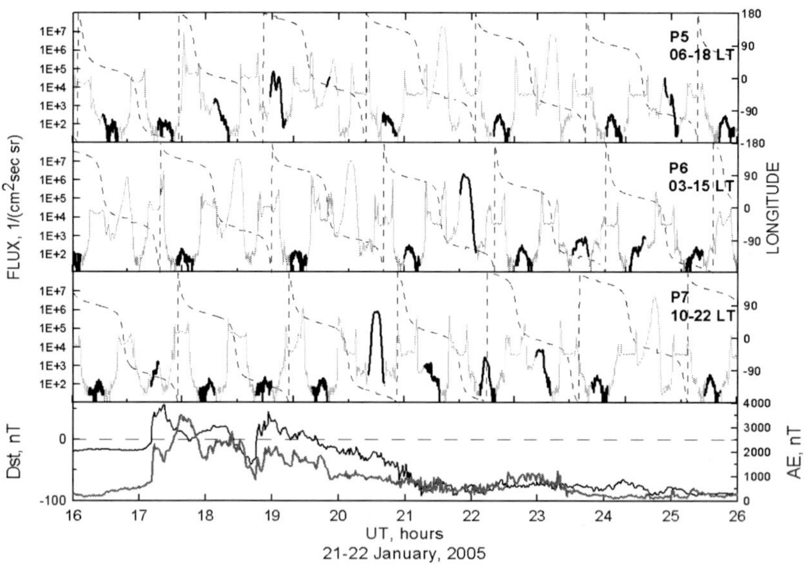

Figure 12. The same as in Figure 10 but for extremely strong compression of the magnetosphere during a moderate storm on 21-22 January 2005.

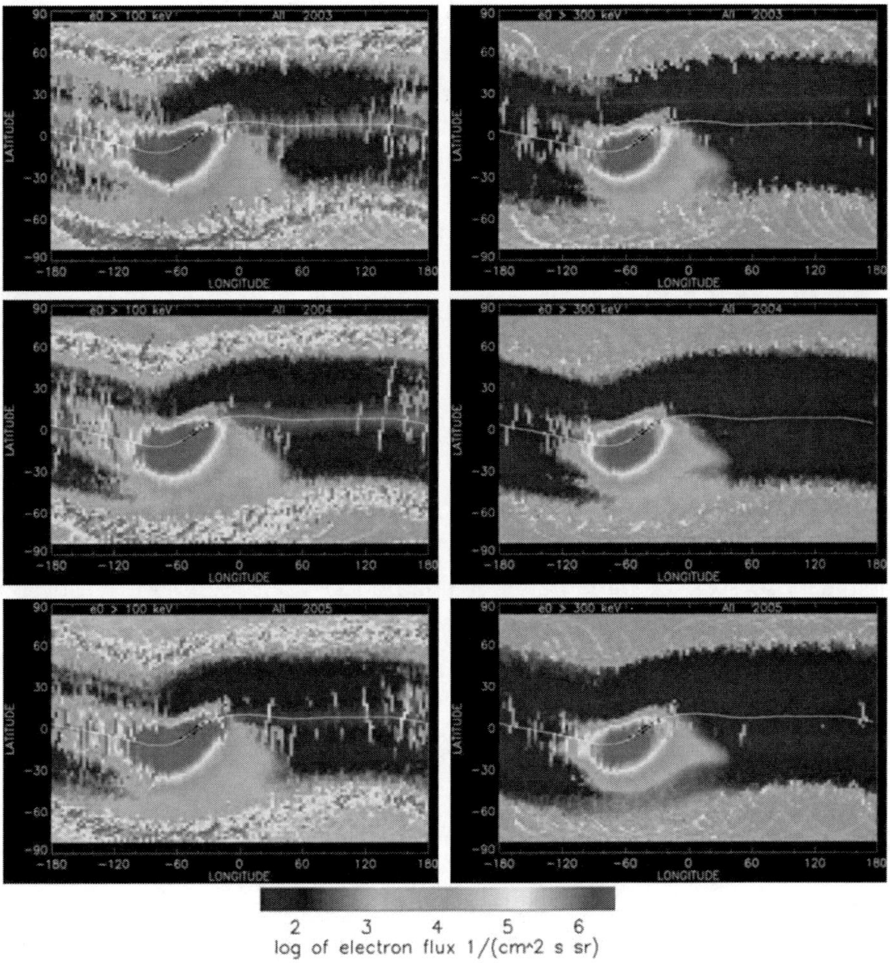

Figure 13. The same as in Figure 9 but for one-year statistics (from top to bottom): in the years 2003, 2004 and 2005.

Figure 13 shows global maps of >100 keV and >300-kev electrons compiled from measurements of the detector pointed to zenith for three years 2003, 2004, and 2005 during the declining phase of solar cycle 23. It is clearly seen from comparison of Figure 13 with Figure 9 that contributions of the presented cases to the corresponding one-year distributions are the biggest, probably because these major storms were either of super-storm series (October 2003 and November 2004) or isolated but the strongest one (May

2005). Also note that FEE enhancements in both energy ranges are moderate ~$10^4 - 10^5$ units during the rest of geomagnetic disturbances.

It is hard to predict the forbidden electron enhancements, because a mechanism of the process is not established yet. First of all, one should determine the source of forbidden electrons. As discussed in paper [Suvorova et al., 2014a], one of most likely candidates is the radiation belt. But the process of particle penetration from the radiation belt to the forbidden zone is still unclear. It might be a fast radial drift in strong electric fields of ~20 mV/m occurred at *L*-shells below 2. However, there is a lack of experimental measurements in this region. Hence, this concern needs further experimental data and modeling (see Garner et al. [2004, 2007]).

Meanwhile, one can expect that if energetic electrons occasionally enhanced in the forbidden zone then duration of the event could be comparable with or even less than the period of azimuthal drift of an energetic electron around the Earth. For example, 100 keV electrons, having the drift period of ~6 hours at a height of ~900 km, can pass from Japan to SAA (~120° in longitudes) for two or three hours. While 300 keV electrons drift eastward approximately 3-4 times faster and spend ~30 minutes for running ~120° separation in longitudes.

Hence, a sporadic enhancement, resulted from a single short-time electron injection, can be observed by several satellites at different locations but within a relatively short period of time. An example of such behavior is a storm on 21-22 January 2005 shown in Figure 12 (see also Suvorova et al. [2014b]). After the substorm expansion, the forbidden electrons were observed by different POES satellites in the longitudinal range from 160°E to 110°W (~90° longitude interval) during three hours.

However, from observations of >100 keV electrons, we found many long-lasting enhancements (see Figures 10 and 11). Two events in the case of 7-8 January 2005 are separated by 8-hours interval. Taking into account the drift period of ~6 hours, we can conclude that those enhancements occurred independently and they were probably driven by two successive substorms revealed in the increases of *AE*. Similar causal relation between the substorms and enhancements can be found in the case of 15 May 2005. Both cases support our suggestion that a strong substorm serves as indicator of a process resulting in FEE enhancements.

We can also suggest that the long-lasting high-energy electron enhancements relate to the long-lasting high auroral activity. We conclude that in these cases a prolonged or multiple triggering resulted in long-lasting high-energy electron enhancements.

Thus, the forbidden electron enhancements relate closely to a geomagnetic activity and, hence, are driven by the variations in solar wind dynamic pressure and interplanetary magnetic field [Suvorova et al., 2013a, 2014a,b].

It is well known, that both northward IMF turning and diminishing of the *By* component can be considered as possible triggers of substorms [e.g., Lyons, 1995; Hsu and McPherron, 2004]. Recently, direct evidences were found that solar wind fluctuations substantially affect global convection leading to a dramatic increase in substorm occurrence even under northward IMF conditions [Lyons et al., 2009].

As a consequence, the storm-time magnetosphere responded to the IMF perturbations in ~10 min that can be seen in the *AL* index peaks and large-scale dipolarizations in the magnetic field observed near geosynchronous orbit at night. The dipolarizations resulted in rapid changes in the configuration of the magnetic field, therefore, inducing the electric fields. As it was concluded on the base of observations and modeling results [e.g., Fok et al., 2001; Nagai et al., 2006; Glocer et al., 2011; Dai et al., 2014], the induced electric fields in its turn are responsible for sudden growth of energetic and relativistic electron fluxes in the radiation belt. For example, during the 22 July 2009 recurrent storm, electron flux increases were preceded by nighttime dipolarizations as observed by the POES and GOES satellites in the outer radiation belt [Glocer et al., 2011].

Moreover, electron fluxes with lower energies were also found to increase in the forbidden zone below the inner radiation belt [Suvorova et al., 2014a]. Thus, storm-time dipolarization events in the magnetospheric magnetic field associated with the solar wind variations result in fast radial transport and energetization of radiation belt electrons. Interplanetary magnetic field fluctuations and associated magnetospheric inductive electric fields play important role in these processes.

Finally, it is important to point out that forbidden electron enhancements in a higher energy range (>300 keV) look mostly like burst events (see Figure 9, right panels). The high-energy electrons are penetration radiation, which can affect satellites causing serious malfunctions. Corresponding applied issues will be considered in the following sections.

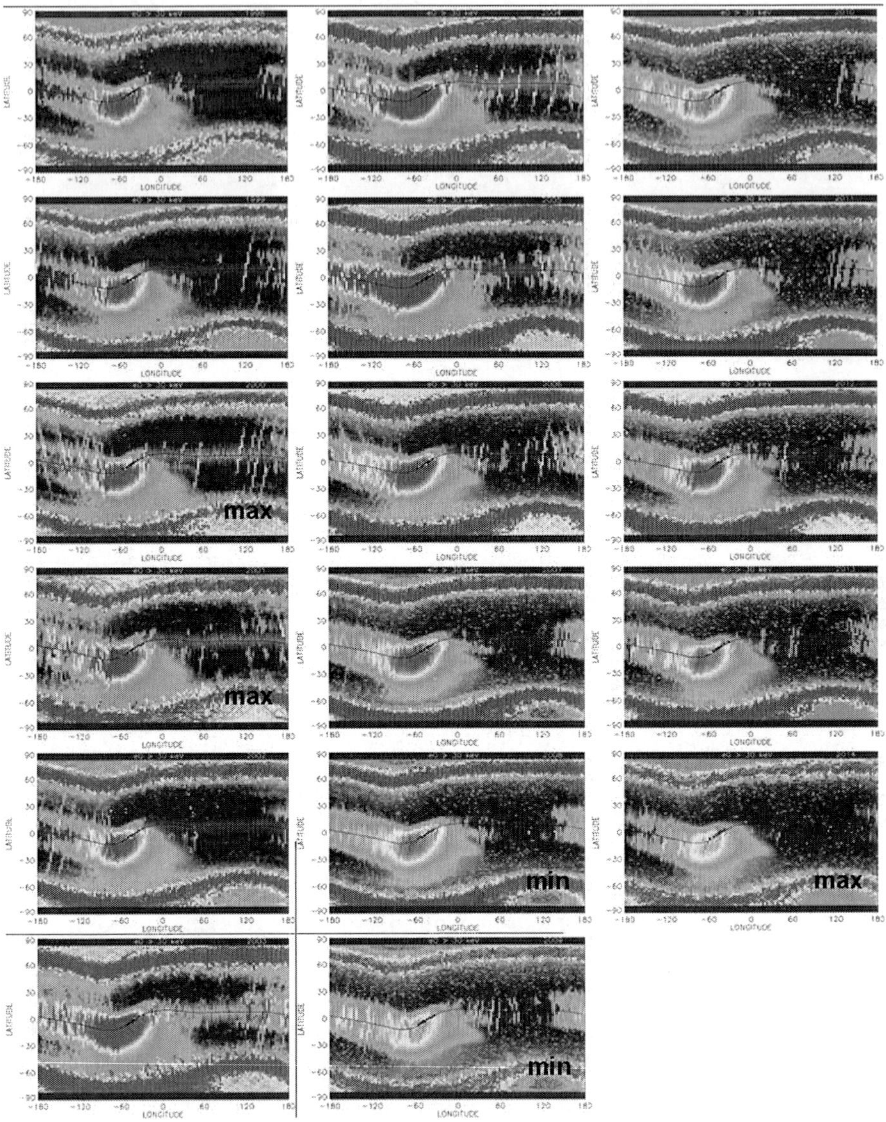

Figure 14. Long-term variation of FEE enhancements of >30 keV electrons. One-year global maps are shown in the 17-year period from 1998 to 2014.

2.6. FEE Statistics and Solar Cycle Variation

We have analyzed FEE events collected in the time interval from 1998 to 2014. For each year, we have compiled a global map, which contains data on the >30 keV electron fluxes from the 0°-detector (Figure 14). This 17-year period includes solar cycles 23 and quite half of 24 that allow us studying a long-term variation of the quasi-trapped energetic electrons in relation with solar activity.

We use sunspot number as a simple and robust parameter for comparisons and ordering the data. Figure 15 shows annually averaged monthly smoothed sunspot numbers in the time interval from 1997 to 2014. The time profile has two prominent maxima of the solar cycles 23 and 24 in 2000-2002 and 2014, respectively, and long minimum in 2008-2009. Several global maps of >30 keV electrons are superimposed on the graph for the following years: 1998, 2000, 2002, 2003, 2005, 2008, 2011, and 2013. These years were chosen for representing different phases of solar cycle: rising, maximum, declining, and minimum.

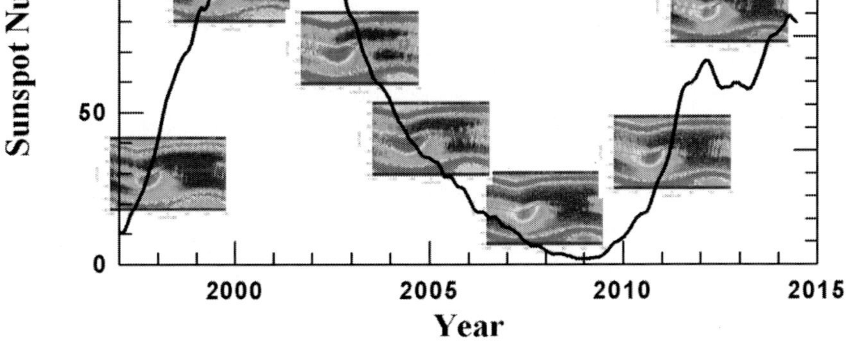

Figure 15. Solar cycle variation of FEE enhancements. The sunspot number and 1-year global maps of >30 keV electrons for 1998, 2000, 2002, 2003, 2005, 2008, 2011, and 2013.

In the first turn, we consider the dynamics of FEE at different phases of the cycle 23, because the cycle 24 is not completed yet. The intensity of FEE fluxes increased from the rising phase to the declining phase, and then it went down during minimum phase (Figure 15).

Vast majority of FEE events from 1998 to 2002 (rising, maximum, declining phases) was of moderate intensity 10^4 units. In the solar maximum (2000 and 2001), there were also several stronger events (10^5 -10^6 units). Moreover during this period, the FEE enhancements were characterized by a double-peak structure over Pacific Ocean and a few events at other longitudes. Note, that the double-peak structure means that enhanced fluxes do not achieve the satellite altitudes at the equator. Contribution of the FEE enhancements within 5-year period around the solar maximum was considerably smaller than that during the remaining 4 years of the declining phase (2003-2006) (Table 1, third row).

During the descending solar activity, the global maps are characterized by abundant representation of the FEE enhancements. The greatest FEE events with intensity of > 10^7 units were observed in 2005. It is well known that during the declining phase of the solar cycle 23, there were many powerful CME structures affecting the Earth's magnetosphere. So, it is quite expected that the substantial abundance of FEE enhancements were observed during that time.

It is surprising that the FEE statistics remains high in the solar minimum (2007-2009) (Table 1). Moreover, maximal intensities are stronger by 1-2 order of magnitudes than those at rising phase (1998 and 1999), beginning of declining phase (2002) and even comparable with those in the solar maximum. Notable features of FEE in the deepest solar minimum (2009) are unexpectedly large number of occurrence and high flux intensities. Note that during the solar minimum, the CIR/HSS-type structures prevailed in the solar wind.

As one can see in Figures 14 and 15, characteristics of FEE events during two strongly different solar cycles also vary in intensity, occurrence and longitudinal extent. From comparison of FEE statistics at rising phases of solar cycles 23 and 24, we find that the weaker solar cycle associates with increasing number of FEE events of moderate intensity without double-peak structure. On the other hand in the maximum of the weaker solar cycle 24, the FEE intensities did not exceed 10^4 units.

3. A Source and Mechanism of FEE Enhancements

3.1. IRB As a Magnetospheric Source of FEE

One might think that a source of enhanced fluxes of the forbidden electrons is not so obvious. Several possibilities can be considered, that is energetic electrons can be locally accelerated or transported to a height of their observations from other plasma domains located below or above such as middle atmosphere/ionosphere and plasmasphere/magnetosphere, in particular ring current and radiation belt. Some of these ideas were discussed previously [e.g., Savenko et al., 1962; Knudsen, 1968; Paulikas, 1975; Asikainen and Mursula, 2005]. Note that before 1980s, studies of the problem were based on data obtained at low heights between 200 and 400 km.

Searching for an answer to the question about the source requires additional information about energetic electrons and fields at other heights. In particular, results of earlier studies of the behavior of trapped electrons at the lower edge of the IRB in response to geomagnetic activity give us very valuable information [Krasovskii et al., 1958, 1961; Imhof and Smith, 1966; Imhof et al., 1967; Imhof et al., 1981; Rosen and Sanders, 1971; Kudela et al., 1992]. It is interesting, that Imhof et al. [1966; 1981], studying high-energy electrons trapped at lower edge of the IRB in energy range 0.07-1.1 MeV, found multiple energy peaks in the electron spectra, which are continually changing in response to geomagnetic activity. They suggested that this was a result of new injections and the peaks might be formed through the earthward transport of electrons. They noted that pronounced peaks were often observed at $L < 1.2$. Rozen and Sanders [1971] found that after periods of enhanced geomagnetic activity, large increases in electron fluxes were observed deep in the IRB and it remained high as long as geomagnetic activity remained enhanced. It is also interesting that Kudela et al. [1992] found a local peak at ~30 keV near the lower edge of the IRB during quiet-times.

Thus we inferred that during geomagnetic disturbances, energetic electrons in the forbidden zone behave in a similar way as at the lower edge of the IRB. The both, dynamics of energetic electrons trapped in the IRB and the forbidden electron enhancements relate closely to the geomagnetic activity and magnetic field dipolarization in the near-Earth region. Hence, they are driven by the variations in solar wind dynamic pressure and interplanetary magnetic field.

Moreover, observations of relatively high energies of FEE (up to 600 keV) do exclude the ionospheric origin and local acceleration by the ionospheric electric field [Dmitriev and Yeh, 2008; Suvorova et al., 2013a]. Hence, it is naturally to suggest the IRB as most likely source of the FEE enhancements [Dmitriev and Yeh, 2008; Suvorova et al., 2012a].

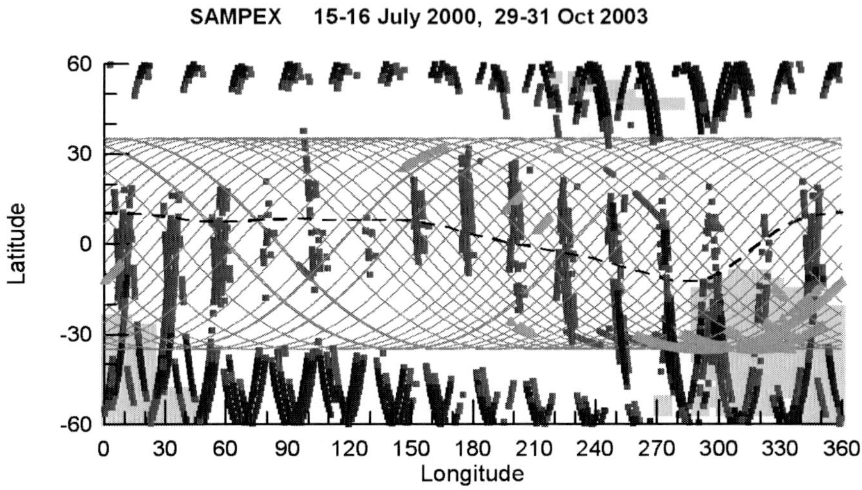

Figure 16. Significant enhancements of the ion density observed by the ROCSAT-1/IPEI in the dayside (red circles) and in the nightside (green triangles) hemispheres, and enhancements of >0.6 MeV electrons and >0.8 MeV protons (squares) observed by the SAMPEX/LEICA at altitudes of ~600 km during the Bastille (15 – 16 July 2000) and Halloween (29 – 31 October 2003) geomagnetic storms. The orbit of ROCSAT-1 during those disturbed periods is shown by gray dotted line. The relative enhancements of particle fluxes of >1.3 and >10 are indicated, respectively, by violet and blue squares. For reference, gray areas depict the regions of strong background particle fluxes (including SAA) with quiet-time intensities of >100 cm^{-2} s^{-1} sr^{-1}. Thick dashed line indicates the geomagnetic equator. Most of the ionospheric ion density enhancements are overlapped with the regions of enhanced fluxes of the energetic particles.

Our suspicion that electrons penetrate into the equatorial ionosphere from higher altitudes, i.e., from the trapped particle zone of the IRB, have been verified with using a satellite located at other heights than POES [Suvorova et al., 2013a, 2014a]. We compare experimental data acquired from NOAA satellites with those from SAMPEX and Japanese satellites SERVIS-1 and GOSAT, which were available for geomagnetic storms on 15 – 16 July 2000 (Bastille storm), 29 – 31 October 2003 (Halloween storm), 9 November 2004 and 22 July 2009. The satellites SERVIS-1 and GOSAT have polar sun-synchronous orbits at altitude of ~1000 km and ~670 km, respectively. The SAMPEX satellite has slightly elliptical near polar orbit at an altitude of ~500 by 680 km. Additionally, we use the data from NOAA/DMSP satellite fleet which measure low-energy electrons at heights of ~850 km. DMSP particle spectrograms are provided online by the Auroral Particle and Imagery Group of JHU/APL at http://sd-www.jhuapl.edu/Aurora/spectrogram/index.htm. Parameters of the orbit for the DMSP satellites are the same as for POES but with covering only the dawn-dusk sector.

Figure 16 shows enhancements of >0.6 MeV electrons and >0.8 MeV protons observed by the LEICA instrument on board the SAMPEX satellite during the Bastille (15 – 16 July 2000) and Halloween (29 – 31 October 2003) geomagnetic storms [Dmitriev and Yeh, 2008]. The fluxes are normalized to the quiet day fluxes. The relative enhancements of particle fluxes of >1.3 and >10 are indicated. For reference, gray areas depict the regions of strong background particle fluxes (including SAA) with quiet-time intensities of >100 units.

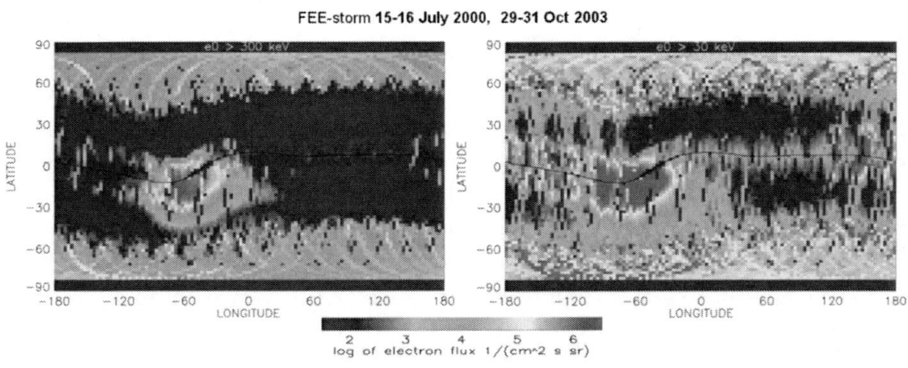

Figure 17. Global maps of FEEs combined for super-storms on 15-16 July 2000 and 29-31 October 2003. FEEs are shown for the energy range (left) >300 keV and (right) >30 keV.

Figure 17 shows POES observations of enhancements of >300 keV electrons (left) and >30 keV (right) during the same two storms. The measurements at the both heights show enhancements at longitudes over Pacific Ocean from ~150°E to ~260°E. The intensity of >300 keV electrons at height of ~850 km increased by 1-3 orders of magnitude, that is higher than at lower height of ~600 km. Hence, the intensity of high-energy electrons increased with altitudes.

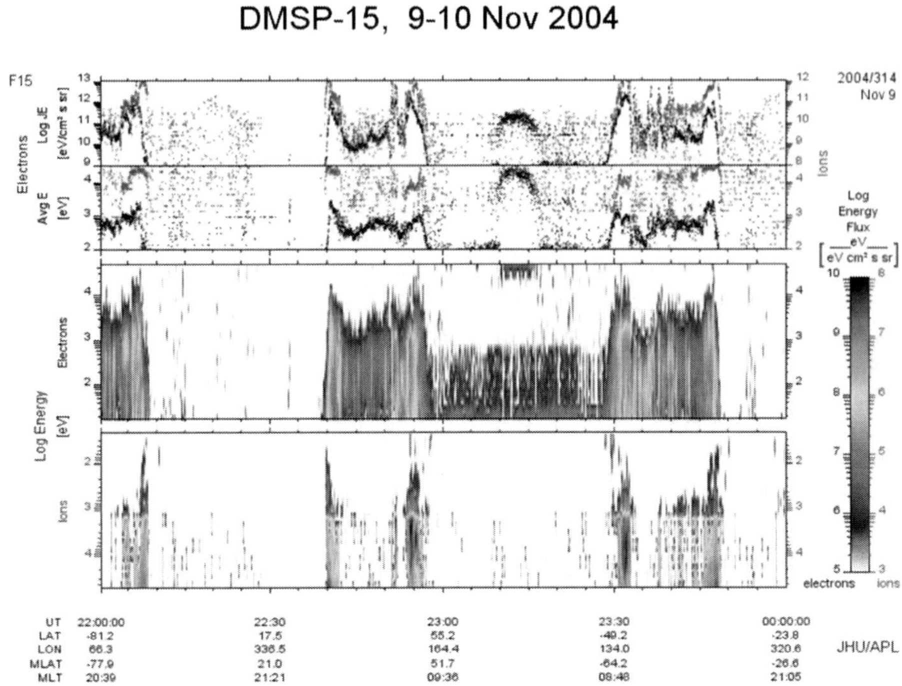

Figure 18. An example of storm-time FEE event in suprathermal electrons (30 eV - 30 keV) measured by DMSP-15 satellite from 22 UT to 24 UT on 9 November 2004 (see text for details).

Figures 18-20 represent the data on electrons from suprathermal to relativistic energies measured during 9 November 2004 storm on board NOAA satellites (POES and DMSP) and SERVIS-1 satellite. Figure 18 shows fluxes of low-energy electrons and ions measured by the DMSP 15 satellite. At around 23 UT, fluxes of electrons in energy ranges 0.03-1 keV and 10-30 keV strongly increased up to 10^7 units at longitude of ~150°E. Measurements in the medium energy range ~30 keV on board DMSP well agree with the fluxes detected by POES (Figure 19, right). Figure 20 shows relative electron fluxes in high-energy channels of 0.3-1.5 MeV and 1.7-3.4 MeV measured by a telescope on board SERVIS-1 satellite at altitudes ~1000 km [Suvorova et al., 2013a]. Strong electron enhancements of $>10^3$ and ~100 units, respectively, in both energy ranges can be found at low *L* shells. From comparison of POES and SERVIS-1 measurements of high-energy electrons (Figure 19, left and Figure 20) one can see that the particle intensity at height of ~850 km increased by ~2 orders of magnitude, that is smaller than that observed at higher altitude of ~1000 km. Hence, the intensity of high-energy electrons increased with the height.

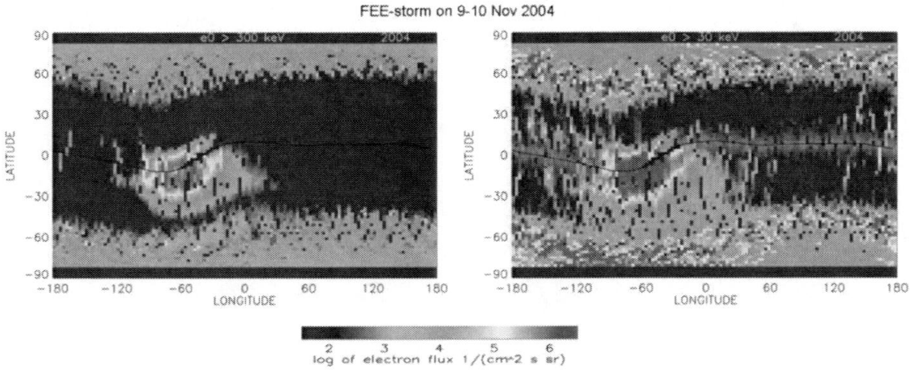

Figure 19. An example of the FEE of medium and high energies: (left) >300 keV and (right) >30 keV during a storm on 9-10 November 2004.

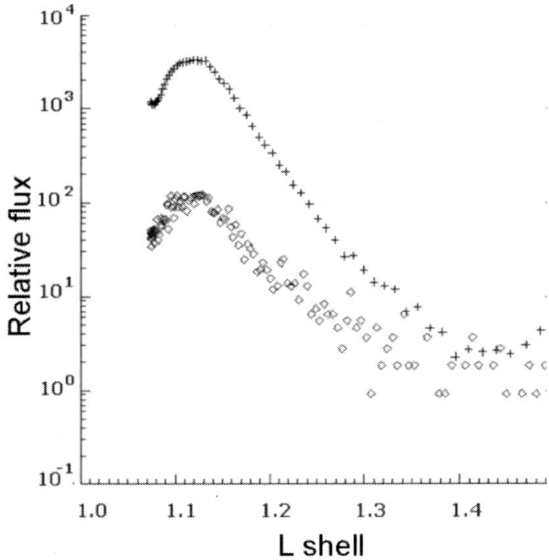

Figure 20. An example of FEE of high and relativistic energies during the storm on 9 - 10 November 2004. *L*-shell profiles of relative electron fluxes measured by SERVIS-1 satellite in energy channel 0.3 - 1.5 MeV (black crosses) and 1.7 - 3.4 MeV (blue diamonds). The fluxes are normalized to those during a quiet day on 8 November 2004.

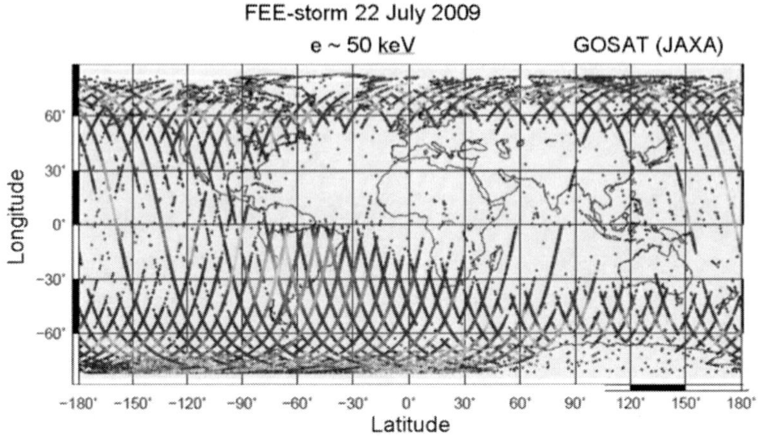

Figure 21. Geographic map of differential electron fluxes in the energy channel 53 - 61 keV measured by GOSAT satellite at ~600 km altitude on 22 July 2009 recurrent magnetic storm.

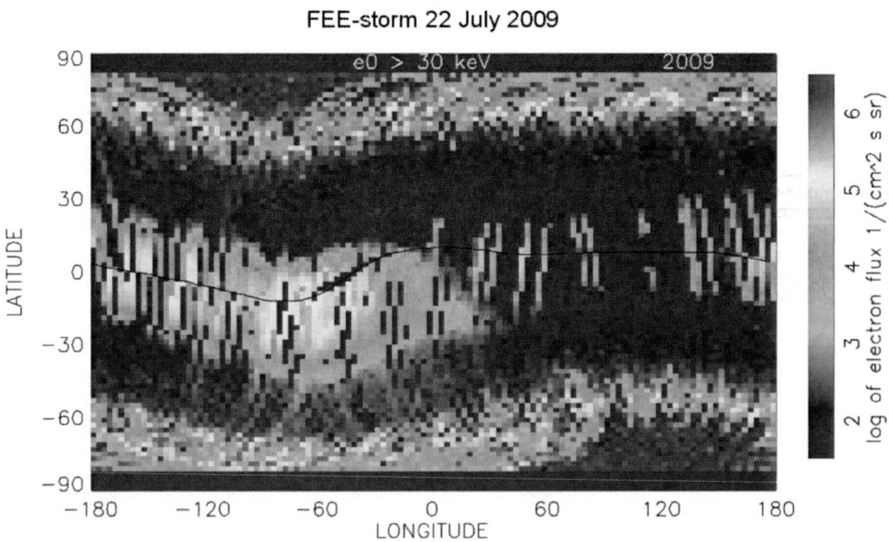

Figure 22. Geographic map of >30 keV FEE enhancements during 22 July 2009 recurrent magnetic storm.

The last example is presented in Figures 21 and 22 (see Suvorova et al. [2014a] for more details). Figure 21 shows global map of differential electron fluxes in the energy channel 53-61 keV. Low-latitudinal enhancements of the quasi-trapped electrons were observed by the GOSAT satellite over Pacific region from 150°E to 120°W. An equatorial peak of differential flux observed for the energy ~50 keV was ~10^7 (cm^2 s sr MeV)$^{-1}$ that corresponds to the integral flux of ~3×10^4 units. At the same time and in the same geographic region, POES observed electron flux of >10^5 units at height of 200 km higher (Figure 22). Hence, we can conclude that the flux of electrons increases with the height, i.e., they have a positive vertical gradient that indicates to the IRB as a source of FEEs. For this particular event, the quasi-trapped electrons were observed in the midnight ionosphere at altitudes of the *F* region. Moreover, the observations of FEEs by the GOSAT satellite are found to be in very good agreement with those observed by the NOAA/POES satellites in the topside ionosphere.

In the examples presented above, we have showed that the fluxes of forbidden energetic electrons have a positive vertical gradient that proves their origin from the IRB.

3.2. Fast Radial Transport as a Mechanism for FEE Enhancements

Questions of a source and mechanism of FEE enhancement are closely coupled. If the IRB is a source of FEEs then it is natural to suggest an earthward radial transport due to $E\times B$ drift as a mechanism of FEE enhancements. Conditions required for the transport of electrons from the IRB to the heights below 1000 km was discussed by Suvorova et al. [2013a; 2014a].

Radial transport can be diffusive or fast (nondiffusive). A well-known mechanism of radial diffusion across the drift shells is very slow because of strong magnetic field at low altitudes. So the slow diffusion results in very weak fluxes of energetic electrons at forbidden drift shells. In contrast, geomagnetic storms, large substorms, and sudden commencements or magnetospheric compressions are accompanied by a very strong penetrating electric field (EF) in the dawn-dusk direction with amplitudes varying from several mV/m to 20-40 mV/m [e.g., Wygant et al., 1994, 1998; Kozelova et al., 2000; Nishimura et al., 2006; 2009; Shinbori et al., 2006; Fejer et al., 2007; Lazutin and Kuznetsov, 2008; Dai et al., 2014]. On the nightside, these substorm-induced and shock-induced electric fields are pointed westward and result in the fast (a few hours) E×B drift of particles across magnetic field lines toward the Earth [e.g., Li et al., 1993; Gabrielse et al., 2012]. Recently, Van

Allen Probe observations found nonlinear EF structures within L shell of ~ 4 – 6, which can be closely associated with magnetic field dipolarizations and enhancements of 1 – 30 keV electron energy flux [Malaspina et al., 2014].

Numerous studies of energetic particle injections into the inner magnetosphere at L~ 3 - 6 associated with intense EFs were reported in literature [e.g., Garner et al., 2004; Lazutin and Kozelova, 2012; Dai et al., 2014]. The studies showed that fast $E\times B$ drift during storms and substorms produces effective transport of radiation belt electrons over a broad energy range in the inner magnetosphere. We suggest that the same process might lead to the radial transport of energetic electrons into forbidden zone below the radiation belt. We have demonstrated that the electrons over a wide energy range from 30 eV to >1.7 MeV appeared at low L-shells so simultaneously that our idea regarding earthward electron transport due to $E\times B$ drift, which is energy independent, was supported.

Concerning the strong electric fields in the innermost magnetosphere, there is an experimental evidence for their generation, particularly at L ~ 2 – 4 by the CRESS satellite. It was shown that measurable electric fields (quasi-static and low frequency) ranging from 1 to 40 mV/m penetrated deeply to the inner magnetosphere in the equatorial plane at least to L = 2 and may be extended beyond [e.g., Wygant et al., 1994; 1998; Burke and Maynard, 2000]. Very strong impulses of the induced EF of ~20 mV/m were revealed in the mid-latitude magnetosphere [e.g., Kozelova et al., 2000]. Likely, similar strong impulses of EF can penetrate to low latitudes at heights of lower edge of the IRB. However, the electric field at L shells below 4 is still a poorly explored subject due to a scarcity of regular measurements. Recently, the Van Allen Probe mission began measurements of electric fields in the innermost part of the magnetosphere.

The measurements at L >4 clearly show that the large-scale dawn-dusk convection electric field and the fluctuating electric fields are significantly enhanced at night side and especially early morning for time periods of an hour or more during major storms [Matsui et al., 2013]. If so, electrons of the RB will suffer $E\times B$ drift toward the Earth. During quiet-times, the energy spectrum of the trapped electron population at L ~ 1.23–1.34 has a local peak at 30 keV [Kudela et al., 1992]. During enhanced geomagnetic activity, these electrons suffer earthward $E\times B$ drift that results in the largest flux enhancements. Then electrons drift eastward along drift shells whose altitude above the Pacific region decreases with increasing longitude (see Figure 2).

The observations convincingly show that the enhanced convective EF results in increases of energetic particle fluxes down to L ~ 1.5 - 3 [e.g., Imhof et al., 1981; Glocer et al., 2011]. These fluxes might be a seed population that could be acted on by strong impulsive EFs related to dipolarization events occurred repeatedly during the storm main and recovery phases. Suvorova et al. [2014a] have showed that electron injections in the forbidden zone and SAA during enhanced geomagnetic activity (*AL/AE* index) on 22 July 2009 are preceded by the dipolarizations in the nighttime magnetosphere. Thus, strong inductive electric fields might transport the seed population from the IRB further earthward that makes possible to penetrate the electrons to ionospheric heights, particularly, in the forbidden zone.

We believe that the observed behavior of low-, medium- and high-energy electrons at $L < 1.1$ is a signature of the rapid earthward transport due to the $E \times B$ drift of radiation belt electrons occurred repeatedly during a storm and/or substorm. Large-scale convective electric fields usually increase during a storm main phase, while strong impulsive electric fields are induced during substorm dipolarization events, i.e., roughly at any time.

4. Impact of FEEs on the Equatorial Ionosphere

At the heights of upper atmosphere, the primary energy of energetic electrons is lost in ionization and excitation of thermospheric species and in production of secondary electron. These processes can alter the ionospheric electron density and chemical composition and can, in turn, influence the chemical reaction rate, and, in particular, the recombination rate (see Danilov and Ivanov-Kholodnyi [1965]; Marov et al. [1996]; Sheehan and St.-Maurice [2004] for a review). At lower heights of the bottom ionosphere, the coupling between the magnetosphere and the ionosphere through precipitating particles from radiation belts and the central plasma sheet strongly controls ionization and conductance in the ionosphere at high latitudes as has been known for some time (see Kelley [2009] for reference).

At heights of the topside ionosphere, these complex processes remain incompletely studied. Since the distribution of electron energy losses in the ionosphere is difficult to determine, information regarding electron losses inside the topside ionosphere are not available yet [Sheehan and St.-Maurice, 2004]. For simplification, one can arbitrarily assumes that electrons lose 100% of their energy in ionization of atmospheric atoms.

Studies of the magnetosphere-ionosphere coupling at low latitudes have begun after revealing FEE enhancements, when it had being shown that FEEs can reach the topside ionosphere and increase ionization at altitudes 300 - 600 km [Dmitriev and Yeh, 2008; Suvorova et al., 2012a]. As discussed above during intense FEE events, quasi-trapped electron fluxes are comparable to the intensity of precipitating electrons within the SAA region and in the auroral zone, where the importance of the ionizing particle effect in the ionosphere is well established. Many FEE enhancements lasted for a long time. Since the FEE fluxes in the *F* region and topside ionosphere were quite intense and, what is more important, long lasting, then their energy deposition in the low-latitude ionosphere can be appreciable.

By assuming that energetic electrons ionize oxygen atoms, which dominate the ionosphere, and accepting additional assumption about the recombination rate at heights 300-600 km, one can estimate the production of ion-electron pairs. Dmitriev et al. [2008] numerically computed ionization losses of energetic electrons bouncing between the top and mirror points. Charged particles spend most of time in the vicinity of mirror points located at lowest heights where the atmospheric density and, thus, the probability of collisions are highest. Therefore, energetic electrons deposit most of the energy near the mirror points. Note, that the efficiency for ionization is partially reduced with height due to decreasing of thermospheric density between 300 and 600 km. By applying the computing technique by Dmitriev et al. [2008] for FEE at energy of 30 keV and at height between 800 km and below 600 km, it was found that within ~2 hours an electron loses the whole energy. This time is comparable to the drift period of several hours. Following this calculation scheme, Suvorova et al. [2013a] have estimated maximal ionizing effect with an excess value of total electron content (TEC). The excess TEC values varied from a few to about hundred TECU (1 TECU = 10^{16} electrons m^{-2}) depending on the energy spectra of electrons. The analysis yields the largest energy deposition at low-to-mid latitudes rather than at the equator, corresponding well with the spatial location of low/mid-latitude positive ionospheric storms [Suvorova et al., 2013a].

Suvorova et al. [2013a] investigated a response of the low latitudinal ionosphere on the great FEE enhancements during CME-driven major storms in time interval from 1998 to 2006. Contribution of ionizing FEE effects to a recurrent storm was analyzed in [Suvorova et al., 2014a]. Ionospheric storm positive phases were analysed using satellite-born and ground-based data. Some illustrations of FEE ionizing effect are presented here, while the quantitative assessments can be found in [Suvorova et al., 2013a, Tables 1 and 2]

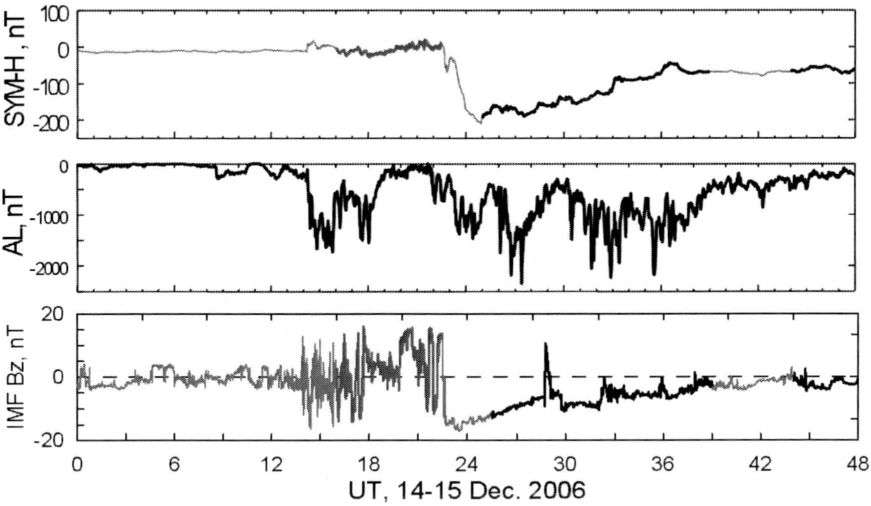

Figure 23. Geomagnetic and solar wind conditions during strong magnetic storm on 14-16 December 2006 (from top to bottom): geomagnetic indices SYM-H and *AL*, IMF *Bz* component measured by the ACE upstream monitor.

4.1. Ionizing Effect during Transient and Recurrent Storms

As example, we analyze effects of storm-time FEEs on 14-15 December 2006. Storm-time dynamics of the FEEs was discussed in section 2.3, Figure 6. Figure 23 shows geomagnetic activity indices (*Dst*/*SYM-H*, *AL*) and IMF *Bz* from 14 to 15 December 2006. The storm initial phase was lasting from ~14 UT until ~23 UT. After the main phase of short duration of ~2 hours, the *Dst*-variation slowly recovered about 2 days.

We compare ionospheric storm positive phases with concomitant FEE enhancements at different phases of the geomagnetic storm. In Figure 6, the first FEE enhancement appeared between 16 and 2230 UT during the initial phase on 14 December. The pre-storm FEE event is characterized by a large intensity up to 10^5 units and duration of 6 hours. The FEE event was preceded by series of substorms (*AL* ~ -1500 nT) between 14 and 18 UT. After 18 UT, the auroral activity decreased because IMF *Bz* turned northward (Figure 23).

Long-lasting ionospheric positive storm together with concomitant FEE enhancements is shown in Figure 24. Ionospheric positive storm is determined as positive residual between storm-time and quiet-time vertical TEC (VTEC) values. VTEC values are provided by the Center for Orbit

Determination (CODE) in the form of 2-hour Global Ionospheric Maps (GIM). The quiet day was 4-5 December 2006.

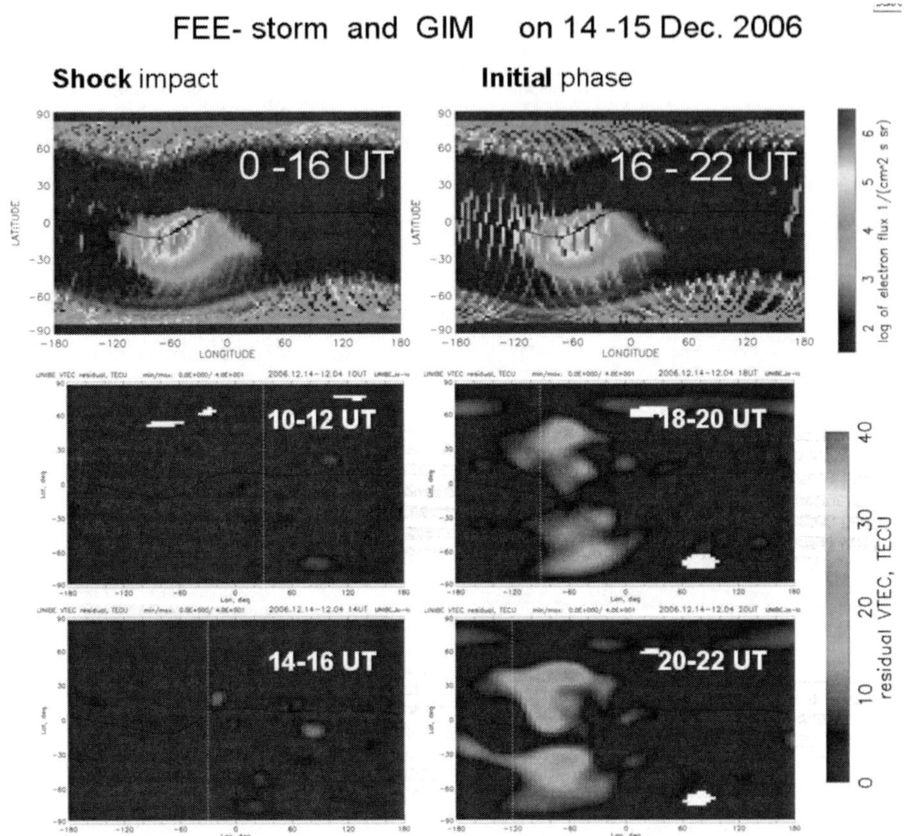

Figure 24. Comparison between FEE enhancements and positive ionospheric storms for different time intervals during the 14-16 December 2006 storm. Left column shows (from top to bottom): geographic map of >30 keV electron fluxes and 2-hour global ionospheric maps (GIM) of residual VTEC during pre-storm period from 0 to 16 UT on 14 December. Right column shows (from top to bottom): geographic map of >30 keV electron fluxes and GIMs during initial phase from 16 to 2230 UT on 14 December. Dotted white line is local noon at the beginning of a 2-hour interval.

Figure 24 compares global maps of FEE and residual VTEC (dVTEC) in time intervals: (left) from 0 to 16 UT including both the pre-storm condition and shock impact at ~14 UT, (right) 16-2230 UT including the greater part of initial phase. Immediately after the shock impact, there was no any increase both in FEE flux and VTEC values until 16 UT. Later at the initial phase, from 16 to 2230 UT, dVTEC values increased up to 20 TECU particularly in the afternoon sector over a wide latitudinal range (±50°) with accompanying by strong FEE enhancements over Pacific Ocean and SAA. Note that the northward IMF direction for ~4 h, from 18 to 2130 UT excludes a priori the prompt penetrating electric field as a general mechanism for the ionospheric positive storm at dayside (see Mendillo [2006] for a review). Other known mechanisms, such as disturbance dynamo and neutral wind, were involved for explanation by Lei et al. [2008b]. During the initial phase, large-amplitude variations of the *Bz* IMF resulted in changing ionospheric electric fields. The daytime positive storm had been modelled by simulating the disturbed electric field and neutral wind circulation. The simulation results showed that the applied model did underestimate the magnitude of the TEC increase at the daytime, especially over SAA (longitudes within ~0°-100°W). Note, however, that in the SAA area the situation is more complicated, because a local electrodynamic mechanism associated with enhanced particle precipitations can operate here [Lin and Yeh, 2005; Foster and Coster, 2007; Dmitriev and Yeh, 2008]. Thus, the enhanced fluxes of energetic electrons can indirectly affect the ionosphere density particularly in SAA. At higher latitudes (> 50°), some contribution from radiation belt particles can be possible.

Figure 25 shows the short main phase and 3-hour interval after the storm maximum. This time interval was comprehensively analyzed by Lei et al. [2008a], Pedatella et al. [2009], and Suvorova et al. [2012, 2013a]. The previous study found quasiperiodic ionospheric oscillations over Japan, the *F* layer uplift by over 100 km, a spotty structure in a large TEC enhancement, ~10 TECU increase in TEC value in the topside ionosphere/plasmasphere at middle latitudes over Pacific Ocean, and a very wide latitudinal extent of positive storm. Here we give an additional comment to some important features.

In Figure 25 (left), one can see a diminishing positive storm, which was not accompanied by any FEE enhancement during the main phase. Apparently, the previously enhanced FEE had time to reach the SAA. But during the storm main phase, caused by southward IMF *Bz*, the fountain effect was intensified by the prompt penetrating electric field. In addition at ~23 UT, the magnetosphere was strongly contracted by a solar wind pressure pulse of > 45 nPa according to measurements from the WIND upstream monitor. Apparently, the solar wind strong impact resulted in a new injection of the inner radiation belt electrons at lower heights. After ~2 hours, the moderately enhanced electron fluxes had observed in the forbidden zone at 6 LT (60°E) and then the new FEE enhancements occupied the dayside from Africa to America. Ionospheric positive storm was intensifying at noon and in the afternoon during the storm maximum. It achieved local maxima of > 40 TECU at the same longitudes, where electron fluxes exceed 10^5-10^6 units. A multi-peak spotty structure, a wide latitudinal extent and other observational facts presented by Suvorova et al., [2013a] are the evidence of FEE ionizing effect in the topside ionosphere. On the other hand, Suvorova et al. [2013a] discussed that the general mechanisms were operating during this time interval, and hence, the direct ionization by energetic electrons can be considered as an additional contribution to the positive storm.

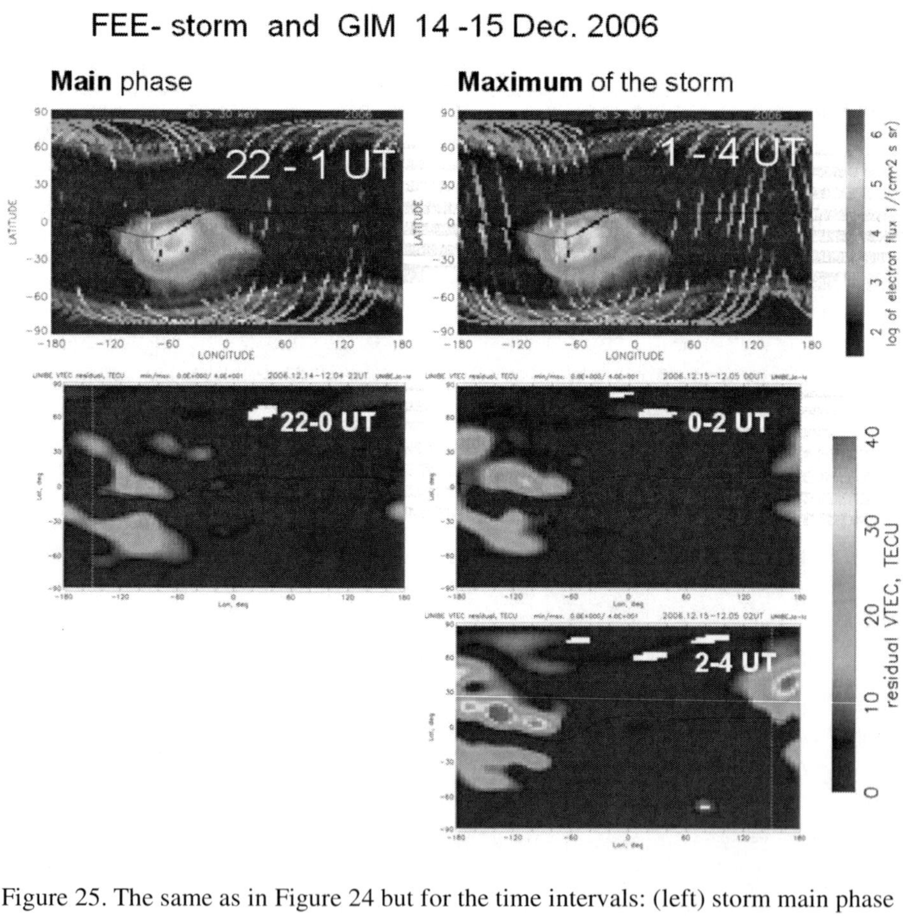

Figure 25. The same as in Figure 24 but for the time intervals: (left) storm main phase from 2230 to 1 UT on 14-15 December, (right) prolonged storm maximum from 1 to 4 UT on 15 December.

However, during the storm recovery phase shown in Figure 26, the general mechanisms can likely be useless for explanation of a long-lasting strong positive storm [cf., Pedatella et al., 2009; Suvorova et al., 2013a]. A simulation of the positive storm by Lei et al. [2008a] failed to reproduce the long-lasting effect over Pacific Ocean region. Previous investigations posed challenges to standard mechanisms identified for this positive storm [e.g., Pedatella et al., 2009].

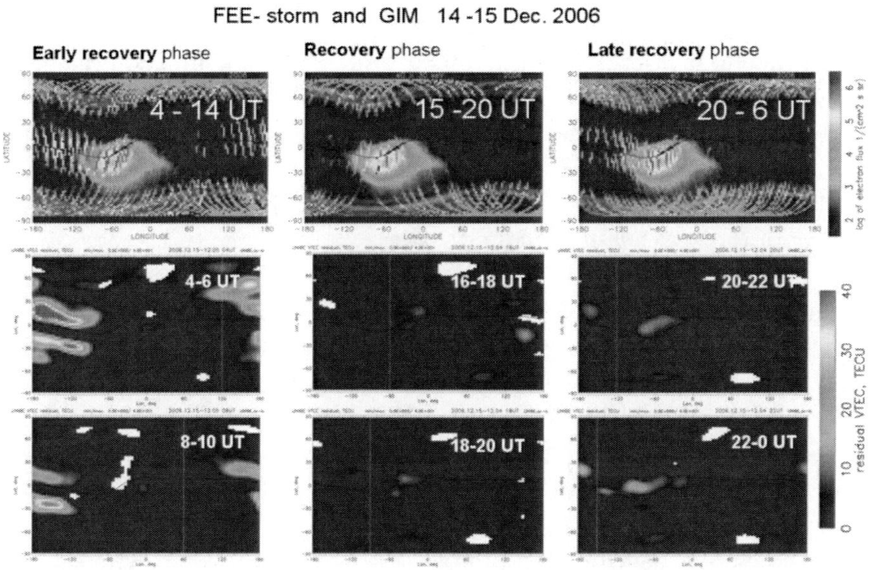

Figure 26. The same as in Figure 24 but for the time intervals: (left column) recovery phase from 4 to 14 UT on 15 December, (central column) recovery phase from 15 to 20 UT on 15 December, and late recovery phase from 20 to 6 UT on 15-16 December.

We divide the 24-hour interval by three parts in accordance with the FEE occurrence. During the first part of the recovery phase from 4 UT to 14 UT (Figure 26, left column), intense electron fluxes were observed within a wide longitudinal range, from Taiwan to America. The positive storm persisted until 12 UT in the same longitudinal area. The maximum of >40 TECU was observed at low latitudes over Pacific Ocean between evening and midnight local times. This strong and prolonged nighttime positive storm is a very unusual feature. From 4 to 6 UT, less intense positive storm (about >30 TECU) was observed at middle latitudes. During the second interval, from 15 UT to 20 UT (Figure 26, second column) both enhancements in FEE and

VTEC disappeared. However, at the late recovery phase from 20 UT 15 December to 6 UT 16 December (Figure 26, right column), the FEE again enhanced, but only at low latitudes. At the same time, one can see a response in increasing of dVTEC up to 10-15 TECU near the equator close to and in the SAA area in the pre-noon and afternoon/evening sectors.

From Figure 24-26, one can see that the most pronounced ionospheric positive storms are concomitant with the FEE enhancements, which repeatedly occurred at all storm phases, from initial to late recovery phase. The ionization effect for the most intense flux of energetic electrons at the recovery phase was estimated to be < 20 TECU [Suvorova et al., 2012a]. Estimations for some super-storms give much larger values [Suvorova et al., 2013a].

Thus, the analysis of the December 2006 transient storm convincingly shows that the direct ionization of the topside ionosphere by quasi-trapped energetic electrons can be considered as an important contribution to positive ionospheric storms at low-to-middle latitudes.

Suvorova et al. [2013a, 2014a, 2015] have presented much more experimental evidences supporting the idea of association of the observed ionospheric storm positive phases with the ionizing effect of the FEE enhancements in the topside ionosphere, that does not depend on type of a geomagnetic storm (transient/major or recurrent/minor). The both phenomena observed simultaneously in the same latitude-longitude regions, repeatedly appearing/disappearing and persisting for approximately similar time periods. Ionizing particle effect has been considered in context of general electrodynamic mechanisms operating at different phase of a magnetic storm and a strong competitive recombination process in SAA and in the nighttime ionosphere. The results allow concluding that ionizing particles of high intensity is an addition source of ionization along with generally accepted mechanisms for the storm-time TEC increase, a positive ionospheric storm. Note that the effects of ionizing particles become mostly pronounced during a recovery phase or initial phase of magnetic storms when other storm-time mechanisms can be ruled out.

4.2. Other Aspects of the Ionosphere Affected by FEE

It was shown that FEEs caused following specific features of enhanced ionization, which are especially difficult for modelling and explanation by standard mechanisms: (1) a wide latitudinal and longitudinal extent, (2) a strong increase of TEC at night, and (3) at late recovery phase, (4) elevation of the *F* region at heights above ~400 km, (5) a multi-spotty structure.

Some other ionospheric effects possibly relate to FEE enhancements being their direct consequences: a formation of the storm-time ionospheric *F3* layer, enhanced ionospheric scintillation activity, and equatorial spread *F* and day-to-day variability. For example, an appearance of the *F3* layer have been reported in literature during July and November 2004 events, and great FEE enhancements also occurred at that time [Suvorova et al., 2013b]. Mentioned above oscillations over Japan in December 2006 storm occurred simultaneously with the FEE enhancements. Coupling of FEE events to these phenomena needs further investigations and testing with direct observations.

5. LEO Spacecraft Anomaly

Knowledge of where and when FEE events will occur can be used not only for applied problems concerning radio signal distortion in the equatorial ionosphere, but also for space weather problems, which are analogous to those impacted on the satellites in the inner magnetosphere: satellite surface/internal charging and electrostatic discharging, solar panel degradation and associated satellite anomalies and failures or errors in on board data processing. Electronic problems, by-turn, may result in mechanical malfunctions, such as an altitude loss and etc.

According to the conventional definition, satellite anomaly is a mission-degrading or mission-terminating event affecting on-orbit operational spacecraft (see Galvan et al. [2014] for a review). Satellite anomalies are actively studied for many years for satellites operated at high Earth orbit, the geosynchronous orbit at an altitude of ~6.6 Re and at low Earth orbits as well [e.g., Garrett, 1981; Gussenhoven et al., 1985; Krupnikov et al., 1999; Degtjarev et al., 2000; Fennell et.al., 2002; Dmitriev et al., 2002; Fennell et al., 2008; Novikov et al., 2008; Bonin et al., 2010; Clilverd et al., 2012]. It was found that energetic particles within a wide energy range from tens of keV to hundreds of MeV can cause different kinds of spacecraft malfunctions (see Kudela, [2013] for references). Particularly, the radiation belts may contribute to surface and internal charging, single event effects and radiation damage.

According to conventional classification, high-energy electrons (~0.5 to ~5 MeV) can cause internal (deep-dielectric) charging while electrons of moderate and lower energy (~10 to ~500 keV) can cause satellite surface charging and electrostatic discharging effects on space systems [e.g., *Baker*,

2001]. Concerning LEO satellite anomaly, much attention is paid to peculiarity of spacecraft charging and electrostatic discharge in the ionospheric/ plasmaspheric plasma environment [e.g., Frooninckx and Sojka, 1992; Bonin et al., 2010] and radiation hazards to spacecraft electronics especially during passing the SAA area [e.g., Belov et al., 2004; Nusinov et al., 2012; Suparta and Gusrizal, 2014].

Hence, we can suppose that enhanced FEE of tens to hundreds of keV energies can also produce serious operational problems of LEO satellites. Here we illustrate this important aspect and present two events of FEE enhancements observed in the same time period as events of satellite anomaly. The information was taken from an anomaly database compiled by Klanowski and publicly available at http://www.sat-index.co.uk/failures/.

First example is a malfunction of the instrument on board the Suzaku satellite just weeks after its launch on 10 July 2005. Suzaku is a Japanese X-ray astronomy satellite at a near-circular orbit at ~570 km altitude with a low inclination angle of 31°. Series of malfunctions with a cooling system of one of the X-ray instruments were occurred from 29 July to 8 August 2005 [Mitsuda et al., 2007].

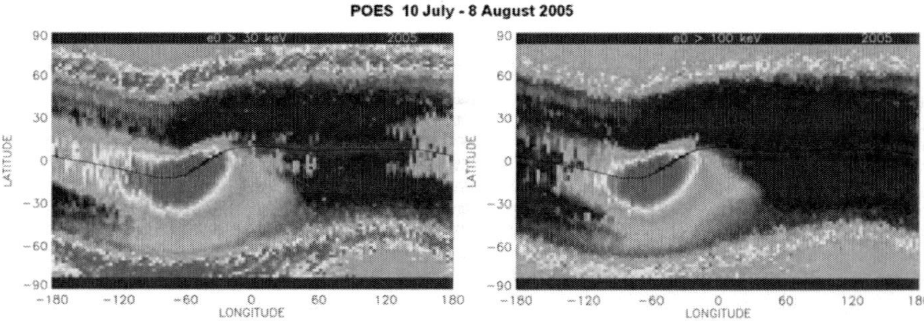

Figure 27. FEE enhancements in energy range (left) >30 keV and (right) >100 keV during time interval from 10 July to 8 August 2005.

Figure 27 shows global maps of FEE of >30 keV and >100 keV energies, compiled from the data from 10 July to 8 August 2005. Geomagnetic activity during these 30 days enhanced periodically (Figure 28). Moderate energy electrons show a large increase with a wide longitudinal extent, while high energy electrons, which are able to penetrate inside the satellite, are characterized by a relatively weak enhancement. Inspecting the interval, we find that FEE of >30 keV energy repeatedly enhanced in response to geomagnetic activity variations, sometimes lasting up to 3 days (10-12 July and 20-22 July). Actually, there was no any day with an absolutely quiet level of FEE fluxes in all longitudinal range, except 8 August. The most intense >30 keV FEE fluxes of 10^5-10^6 units were observed on 11, 12, 18, 20, 21, and 28 July. As example, global maps for 12 July, 18 July and 28 July (the day before the first malfunction on board Suzaku) are shown in Figure 29.

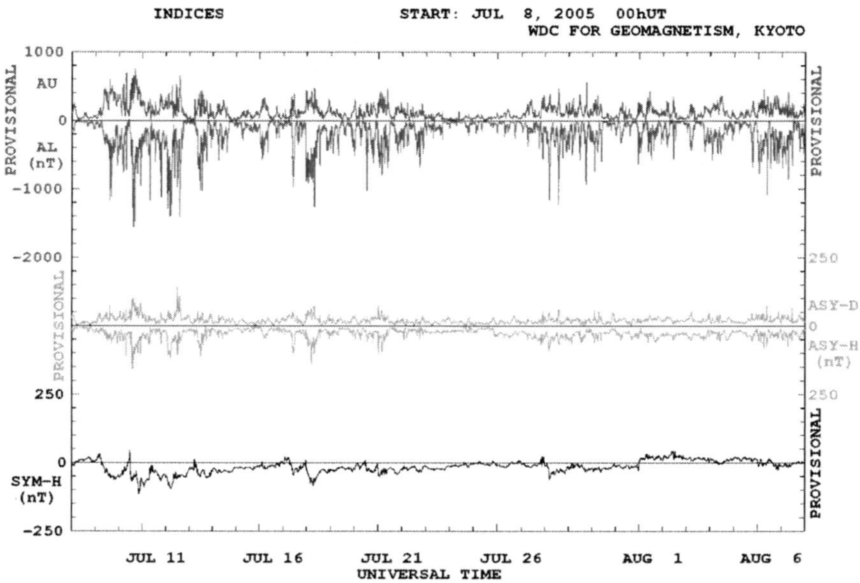

Figure 28. Geomagnetic activity indices in the period from 10 July to 8 August 2005 (from top to bottom): auroral *AU* and *AL* indices, ASY-D, ASY-H and SYM-H indices.

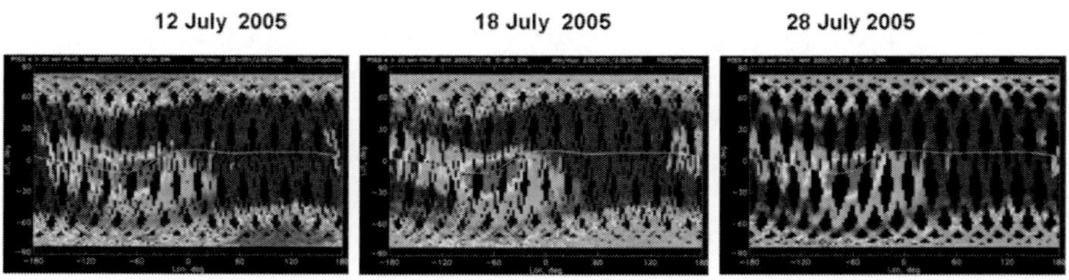

Figure 29. Strong enhancements of >30 keV FEE on 12, 18 and 28 July 2005.

Another example is serious problems with the attitude control system and solar panels of the X-ray observatory Astro-D. This Japanese satellite was launched in 1993 at altitude of 500-600 km with an inclination angle of 31° [http://space.skyrocket.de/doc_sdat/astro-d.htm]. The satellite anomaly was recorded on 15 July 2000. Although only one POES satellite operated at that time, it was lucky to detect FEE enhancements in three energy ranges >30, >100, >300 keV at ~21 UT (Figure 30). Time of electron injection cannot be determined accurately with one satellite. The enhanced low-to-high energy electrons can be considered as potential causes of the satellite anomaly, including the surface/internal charging.

Although these two anomaly events had got already an acceptable explanation, we believe that an analysis of potential impact of associated FEE enhancements is absolutely essential.

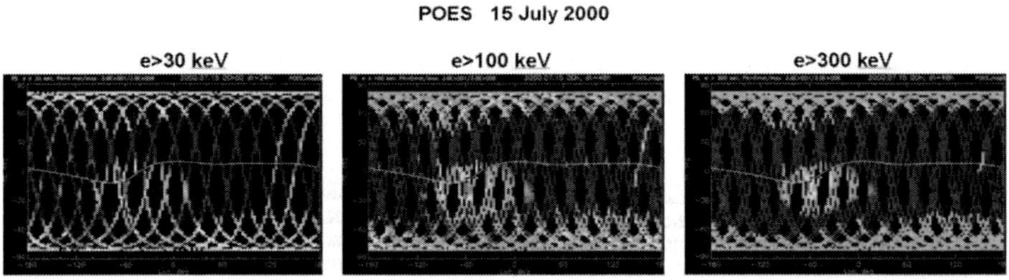

Figure 30. FEE enhancement in energy ranges: (left) >30 keV, (middle) >100 keV and (right) >300 keV on 15 July 2000.

6. X-Ray Contamination to Astrophysical Observations

It is well known that high-energy electrons are able to penetrate the spacecraft surfaces and exterior. In this connection, another important aspect of the FEE radiation effects is the observational astrophysics, because high-energy electrons cause a background noise in X-ray measurements (see Dean et al., [1991] for a review). Many experimental studies were performed in the past to find out a role of the magnetospheric electrons contaminating astrophysical observations at low altitudes. Here we give a short review of these studies.

Astrophysical sources of X-ray and γ-ray emission exist throughout the Universe including galaxies, the solar system and, particularly, the solar corona. To distinguish the emission of a cosmic object from background noise, which is higher in most of cases, is a task of high priority in the observational X-ray and γ-ray astronomy. Identification of all individual sources contributed to the background noise allows improving the techniques for background reduction. Thorough studies show that, particularly, in the X-ray and low-energy γ-ray ranges (from ~100 eV to ~300 keV), the background noise is extremely complex in its origin [Smith et al., 2009; Takahashi et al., 2012]. Nevertheless, two major components in the total background should be distinguished in the first turn: one of astrophysical origin, such as cosmic background, and another of local intrinsic instrumental origin, and besides the last affects on measurement accuracy of the first. As it was recently pointed out, the most important effect limiting the accuracy of cosmic X-ray background measurements is related to the intrinsic background variation in detectors [Gruber et al., 1999; Bisnovatyi-Kogan and Pozanenko, 2011, Revnivtsev et al., 2012].

Astronomical observations are performed on board high- and low-orbiting spacecrafts or on space platforms, in the Earth upper atmosphere, magnetosphere and interplanetary space. Note that orbit choice is one of the important factors affecting the sensitivity of the instruments operated under conditions of low signal-to-noise ratio. In either case the instruments encounter an intense radiation environment, so the background noise is estimated in a specified radiation environment. The internal instrumental noise is caused by the interactions of energetic charged particles with the spacecraft and detectors. Either by direct penetration or by secondary radiations produced in the payload materials, photon detectors may at times give spurious responses, particularly if the "background" radiations are nonsteady [Seward et al., 1974].

Vast majority of astronomy satellites are low-orbiting, i.e., they pass through zones of enhanced particle radiations, the Earth's auroral zone and outer RB at high latitudes and also SAA at low latitudes. Knowledge of contaminating background radiation of magnetospheric origin is accumulated since 60s and the influence of RB has been well investigated [Dean et al., 1991; Cladis and Dessler, 1961; Holt, 1974; Lum et al., 1997]. Particularly, it was shown that detectors of X- and γ-ray are subject to in-orbit enhanced background noise caused by the magnetospheric high-energy electrons [e.g., Feldman et al., 1996; Jayanthi, 1997; Bucik et al., 2002, 2005; Sarcar et al., 2011; Gusev et al., 2012].

To minimize the charge particle contamination from RB, most astronomy experiments are conducted at low inclination orbits, for example RHESII mission [Lin et al., 2002]. It is considered that the equatorial region away from the SAA is characterised by low background, so measurements made during passages of the spacecraft through the equator-to-low latitudinal region are particularly valuable to high-energy astrophysics. However, even during the early 1970s, the X-ray astronomers revealed unexpected electron-induced contaminant at relatively low latitudes as well, which was a few times higher than the cosmic X-ray background [Seward et al., 1974; Neighbours and Clark, 1974]. They found several events when flux intensity of electrons with energy of tens of keV was as large as ~10^3 el/(cm^2 s sr)$^{-1}$, exceeding quiet level by 2 orders of magnitude. Note that it is still much lower than in the radiation belt and auroral zones [see Bucik et al., 2005].

In addition to astrophysical measurements, ionospheric and atmospheric studies [Antonova and Ivanov-Kholodny, 1961; Antonova and Kazarchevskaya, 1965; Knudsen and Sharp, 1968; Doherty, 1971; Potemra and Rosenberg, 1973; Luhmann and Blake, 1978; Dmitriev and H.-C. Yeh, 2008; Dmitriev et al., 2011; Bobik et al., 2013] and satellite data failures studies mentioned above also found several effects suggesting that electron impact is important factor at low and middle latitudes. That is, more importantly the occasional increases of electron flux below the radiation belt were discovered even earlier in direct satellite-borne measurements [Krasovskii et al., 1958; Savenko et al., 1962] and then corroborated in several studies [Heikkila, 1971; Winningham, 1974; Goldberg, 1974, Lieu et al., 1988]. They reported about sporadic fluxes of very high intensity, which was comparable with the auroral precipitation. However, the direct observations of sporadic events caused strong argument due to a doubt about validity of measured high intensity (see review by Paulikas [1975]). As a result of this, despite importance of low-latitude measurements of electron enhancements recognized earlier [Holt, 1974], further investigation of this phenomenon was not carried out.

Our results, presented in the previous sections, are based on long-term statistics of low-orbiting satellite observations of sporadic and long-lasting enhancements of high-energy (up to >300 keV) electrons in the forbidden zone that allows us concluding that the FFE contamination to high-energy astrophysical and solar hard radiation measurements might be potentially a very important problem.

We have demonstrated that 100-300 keV electron fluxes significantly exceeded a quiet level during major geomagnetic storms and under strong compressions of the magnetosphere as well. Hence, enhanced fluxes of the forbidden high-energy electrons can significantly contaminate to X-ray measurements at low-latitude and low-altitude orbit.

Finally, it is important to point out that forbidden electron enhancements in a higher energy range (>300 keV) look mostly like burst events (see Figure 9, right panels; Figures 13, 19, 27, 30). Moreover, they can be observed at any local time, both nighttime and sunlit side of the near-Earth orbit. Such feature makes it difficult to distinguish the astrophysical X-ray source (solar, cosmic or galactic) from the high-energy electron-induced events. The forbidden high-energy electrons will increase detector noise, which may mask the original source signal.

CONCLUSION

The focus of the present research is on dynamics of the near-Earth radiation during geomagnetic storms and substorms. Comprehensive analysis of large statistics on energetic electrons measured by the fleet of the NOAA/POES satellites revealed a very interesting phenomenon in the topside ionosphere at altitude of < 1000 km that is significant enhancements of the forbidden energetic electrons (FEE). We present results of the study of FEE during major and recurrent magnetic storms in solar cycles 23 and 24, which was conducted on the base of experimental data acquired from 1998 to 2014.

A source and mechanism of FEE enhancement are suggested and discussed. We present several observational facts that prove their origin from the IRB. The observed specific behavior of low-, medium- and high-energy electrons at $L < 1.1$ allows us concluding that it is a signature of the rapid earthward transport due to the $E \times B$ drift of radiation belt electrons, which can occur repeatedly during a storm and/or substorm. The observations convincingly show that the large-scale convective electric field, which usually enhances during a storm main phase, results in increases of energetic particle fluxes in the inner radiation belt. These fluxes might be a seed population that could be acted on by strong impulsive electric fields, which are induced during substorm dipolarization events, i.e., roughly at any time. Thus, strong inductive electric fields might transport the seed population from the IRB further earthward that makes possible to penetrate the electrons to ionospheric heights, particularly, in the forbidden zone.

Knowledge of where and when such environmental events will occur can be used in both space physics and applied researches. We focus on the following radiation aspects of the electron enhancement problem with a short review for each item: 1) ionization impact of 10-30 keV electrons in the ionosphere; 2) possible effects of energetic electrons in satellite anomalies attributed to low-orbit radiation environment; 3) X-ray and gamma-ray contaminations by 100-300 keV electrons to astrophysical observation on board low-orbiting satellites. Coupling of FEE events to these phenomena needs further investigations and testing with direct observations.

ACKNOWLEDGMENTS

The authors thank a team of NOAA's Polar Orbiting Environmental Satellites for providing experimental data about energetic particles and Kyoto World Data Center for Geomagnetism (http://swdcwww.kugi.kyoto-u.ac.jp/index.html) for providing geomagnetic indices. The Global Ionosphere Maps/VTEC Data are produced by European Data Center (http://www.aiub.unibe.ch/ionosphere). Authors are grateful to N. Hasebe from Waseda University and H. Matsumoto from JAXA, Japan, for the SERVIS-1 and GOSAT data. The ACE solar wind data were provided by N. Ness and D.J. McComas through the CDAWeb website. The DMSP particle spectrograms are provided by P. Newell and T. Sotirelis through the JHU/APL website. We appreciate very much to K. Kudela for refereeing the manuscript. This work was supported by grants MOST-103-2811-M-008-050, MOST-103-2111-M-008-016- from the Ministry of Science and Technology of Taiwan and partially by grant NSC103-2923-M-006-002-MY3/14-05-92002HHC_a of Taiwan-Russia Research Cooperation. The work of A.V. Dmitriev was supported by the Project-Oriented Research Program of National Central University of Taiwan.

REFERENCES

Antonova, L. A. & Ivanov-Kholodny, G. S. (1961). Corpuscular hypothesis for the ionization of the night ionosphere, *Geomagn. Aeron.*, *1*(2), 164-173.

Antonova, L. A. & Kazarchevskaya, T. V. (1965). Measurements of soft electron streams in the upper atmosphere at altitude to 500 km, *Space Research, Transactions of the All-Union Conference on Space Physics*, ed. by G. A. Skuridin et al., Science Publishing House, Moscow, June 10-16, 1965.

Asikainen, T. & Mursula, K. (2005). Filling the South Atlantic anomaly by energetic electrons during a great magnetic storm, *Geophys. Res. Lett.*, *32*, L16102, doi:10.1029/ 2005GL023634.

Baker, D. N. (2001). Satellite Anomalies due to Space Storms, in *Space Storms and Space Weather Hazards, NATO ASI Ser.*, vol. *38*, edited by I. A. Daglis, 79-102, Kluwer Acad., Netherlands.

Belov, A., Dorman, L., Iucci, N., et al. (2004). The relation of high- and low-orbit satellite anomalies to different geophysical parameters, in *Effects of Space Weather on Technology Infrastructure, NATO ASI Ser.*, vol. *176*, edited by I. A. Daglis, pp. 147-164, Kluwer Acad., Dordrecht.

Bisnovatyi-Kogan, G. S. & Pozanenko, A. S. (2011). About the measurements of the hard X-ray background, *Astrophys. Space Sci.*, *332*, 57–63, doi: 10.1007/s10509-010-0485-9, 2011.

Bobik, P., Putis, M., Bertaina, M., et al. (2013). UV night background estimation in South Atlantic Anomaly, *Proceedings of the 33rd International Cosmic Ray Conference,* (ICRC). Rio de Janeiro, Brazil, 2-9 July 2013.

Bonin, G., Orr, N., Zee, R. & Cain, J. (2010). Solar array arcing mitigation for polar low-earth orbit spacecraft, in: *Proceedings of 24th Annual Conference on Small Satellites*, 9012 Aug. 2010, Logan, USA

Bucik, R., Kudela, K., Dmitriev, A. V., Kuznetsov, S. N., Myagkova, I. N. & Ryumin, S. P. (2002). Spatial distribution of low energy gamma-rays associated with trapped particles, *Adv. Space Res.*, *30*, 12, 2843-2848.

Bucik, R., Kudela, K., Dmitriev, A. V., Kuznetsov, S. N., Myagkova, I. N. & Ryumin, S. P. (2005). Review of electron fluxes within the local drift loss cone: Measurements on CORONAS-I, *Adv. Space Res.*, *36*, 10, 1979-1983.

Burke, W. J. & Maynard, N. C. (2000). Satellite observations of electric fields in the inner magnetosphere and their effects in the mid-to-low latitude ionosphere, *IEEE Transactions on Plasma Science*, *28*(6). 1903-1911

Cladis, J. B. & Dessler, A. J. (1961). X rays from Van Allen belt electrons, *J. Geophys. Res.*, *66*, 2, 343-350, 1961.

Clilverd, M. A., Rodger, C. J., Danskin, D., et al. (2012). Energetic particle injection, acceleration, and loss during the geomagnetic disturbances which upset Galaxy 15, *J. Geophys. Res.*, *117*, doi: 10.1029/2012JA018175.

Dai, L., Wygant, J. R., Cattell, C. A., et al. (2014). Evidence for injection of relativistic electrons into the Earth's outer radiation belt via intense substorm electric fields, *Geophys. Res. Lett.*, *41*, 1133–1141, doi:10.1002/2014GL059228.

Danilov, A. D. & Ivanov-Kholodnyi, G. S. (1965). Research on ion-molecule reactions and dissociative recombination in the upper atmosphere and in the laboratory, *Sov. Phys. Usp.*, *8*, 92-116.

Dean, A. F., Lei, F. & Knight, P. J. (1991). Background in space-borne low-energy γ-ray telescopes, *Space Sci. Rev.*, *57*, 109-186.

Degtjarev, V. I., Popov, G. V. & Johnstone, A. D. (2000). Solar wind control of spacecraft charging conditions in geostationary orbit during magnetic storms, *Adv. Space Res*, *26*(1). 37 – 40.

Dmitriev, A. V. & Yeh, H.-C. (2008). Storm-time ionization enhancements at the topside low-latitude ionosphere, *Ann. Geophys.*, *26*, 867-876, doi:10.5194/angeo-26-867-2008.

Dmitriev, A. V., Guilfanov, I. I. & Panasyuk, M. I. (2002). Data failures in the "Riabina-2" experiment on MIR orbital station, *Rad. Meas.*, *35*, 499-504.

Dmitriev, A. V., Tsai, L.-C., Yeh, H.-C. & Chang, C.-C. (2008). COSMIC/FORMOSAT-3 tomography of SEP ionization in the polar cap, *Geophys. Res. Lett.*, *35*, L22108, doi:10.1029/2008GL036146.

Dmitriev, A. V., Yeh, H.-C., Panasyuk, M. I., et al. (2011). Latitudinal profile of UV nightglow and electron precipitations, *Planetary and Space Science*, *59*, 733-740, 10.1016/j.pss. 2011.02.010.

Dmitriev, A. V., Huang, C.-M., Brahmanandam, P. S., Chang, L. C., Chen, K.-T. & Tsai, L.-C. (2013). Longitudinal variations of positive dayside ionospheric storms related to recurrent geomagnetic storms, *J. Geophys. Res.*, *118*, doi:10.1002/jgra.50575.

Dmitriev, A. V., Suvorova, J.-K., Chao, et al. (2014). Anomalous dynamics of the extremely compressed magnetosphere during 21 January 2005 magnetic storm, *J. Geophys. Res. Space Physics*, *119*, 877–896, doi:10.1002/ 2013JA019534.

Doherty, R. H. (1971). Observations suggesting particle precipitation at latitudes below 40N, *Radio Science*, *6*(6), 639-646.

Evans, D. S. & Greer, M. S. (2004). Polar Orbiting Environmental Satellite Space Environment Monitor: 2. Instrument descriptions and archive data documentation, Tech. Memo. version 1.4, NOAA Space Environ. Lab., Boulder, Colo.

Fejer, B. G., et al. (2007). Equatorial ionospheric electric fields during the November 2004 magnetic storm, *J. Geophys. Res.*, *112*, A10304, doi:10.1029/2007JA012376.

Feldman, W. C., Symbalisty, E. M. D. & Roussel-Dupre, R. A. (1996). Hard X-ray survey of energetic electrons from low-Earth orbit, *J. Geophys. Res.*, *101*, A3, 5195–5209.

Fennell, J. F., Roeder, J. L., Berg, G. A. & Elsen, R. K. (2008). HEO satellite frame and different charging and SCATHA low-level frame charging, *IEEE Trans. on Plasma Science*, *36*, 2271-2279.

Fok, M.-C., Moore, T. E. & Spjeldvik, W. N. (2001). Rapid enhancement of radiation belt electron fluxes due to substorm dipolarization of the geomagnetic field, *J. Geophys. Res.*, *106*, 3873–3881, doi:10.1029/2000JA000150.

Foster, J. C. & Coster, A. J. (2007). Conjugate localized enhancement of total electron content at low latitudes in the American sector, *J. Atmos. Sol. Terr. Phys.*, *69*, 1241-1252.

Frooninckx, T. B. & Sojka, J. J. (1992). Solar cycle dependence of spacecraft charging on low earth orbit, *J. Geophys. Res.*, *97*, 2985-2996.

Gabrielse, C., Angelopoulos, V., Runov, A. & Turner, D. L. (2012). The effects of transient, localized electric fields on equatorial electron acceleration and transport toward the inner magnetosphere, *J. Geophys. Res.*, *117*, A10213, doi:10.1029/2012JA017873.

Galvan, D. A., Hemenway, B., Welser, W. & Baiocchi, D. (2014). Satellite anomalies. Benefits of a centralized anomaly database and methods for securely sharing information among satellite operators, Rand Co., Santa Monica.

Garner, T. W., Wolf, R. A., Spiro, R. W., et al. (2004). Magnetospheric electric fields and plasma sheet injection to low L-shells during the 4-5 June 1991 magnetic storm: Comparison between the Rice Convection Model and observations, *J. Geophys. Res.*, *109*, A02214.

Garner, T. W., Growley, G. & Wolf, R. A. (2007). The impact of stormtime changes in the non-auroral conductance upon the low- and mid-latitude electric foeld, *J. Atmos. Sol.-Terr. Phys.*, *69*, 1200-1212.

Garrett, H. B. (1981). The charging of spacecraft surface, *Rev. Geophys. Space Phys.*, *9*, 577-616.

Glocer, A., Fok, M.-C., Nagai, T., Tóth, G., Guild, T. & Blake, J. (2011). Rapid rebuilding of the outer radiation belt, *J. Geophys. Res.*, *116*, A09213, doi:10.1029/2011JA016516, 2011.

Goldberg, R. A. (1974). Rocket observation of soft energetic particles at the magnetic equator, *J. Geophys. Res.*, *79*, 5299–5303.

Gruber, D. E., Matteson, J. L., Peterson, L. E. & Jung, G. V. (1999). The spectrum of diffuse cosmic hard X-rays measured with HEAO 1, *The Astrophys. J.*, *520*, 124-129.

Gusev, A., Martin, I. & Pugacheva, G. (2012). The soft X-ray emission of nocturnal atmosphere during the descending phase of the 23rd solar cycle, *Sun and Geosphere*, *7*, 2, 127-131.

Gussenhoven, M. S., Hardy, D. A. & Rich, F., et al. (1985). High-level spacecraft charging in the low-altitude polar auroral environment, *J. Geophys. Res.*, *90*, 11009-11023.

Heikkila, W. J. (1971). Soft particle fluxes near the equator, *J. Geophys. Res.*, *76*, 1076–1078.

Holt, S. S. (Ed.), (1974). *Proceedings of the Workshop on Electron Contamination in X-Ray Astronomy Experiments*, Washington, 26 April 1974, NASA GSFC Rep. X-661-74-130.

Hsu, T.-S. & McPherron, R. L. (2004). Average characteristics of triggered and nontriggered substorms, *J. Geophys. Res.*, *109*, A07208, doi:10.1029/2003JA009933.

Imhof, W. L. & Smith, R. V. (1966). The behavior of trapped electrons and protons at the lower edge of the inner radiation belt, *J. Geophys. Res.*, *71*, 4157-4171.

Imhof, W. L., Reagan, J. B. & Smith, R. V. (1967). Long-term study of electrons trapped on low L shells, *J. Geophys. Res.*, *72*, 2371-2377.

Imhof, W. L., Gaines, E. E. & Reagan, J. B. (1981). High-resolution spectral features observed in the inner radiation belt trapped electron population, *J. Geophys. Res.*, *86*, 2341-2347.

Imhof, W. L., Reagan, J. B. & Smith, R. V. (1967). Long-term study of electrons trapped on low L shells, *J. Geophys. Res.*, *72*(9). 2371–2377, doi:10.1029/JZ072i009p02371

Jayanthi, U. B., Pereira, M. G., Martin, I. M., et al. (1997). Electron precipitation associated with geomagnetic activity: Balloon observation of X ray flux in South Atlantic anomaly, *J. Geophys. Res.*, *102*, A11, 24069-24073.

Kelley, M. C. (2009). The Earth's Ionosphere: Plasma Physics and Electrodynamics. *International Geophysics Series*, vol. *96*, Academic Press, p.556.

Klanowski, P. C. (2012). Satellite News Digest: Satellite Outages and Failures, web pages, http://www.sat-index.co.uk/failures/.

Kozelova, T. V., et al. (2000). Substorm large impulsive electric fields observed by CRESS, *in Proc. 5th International Conference on Substorms*, St. Petersburg, Russia, 16-20 May (ESA SP-443, July 2000).

Knudsen, W. C. (1968). Geographic distribution of F-region electrons with about 10-eV energy, *J. Geophys. Res.*, *73*, 3, 841-856.

Knudsen, W. C. & Sharp, G. W. (1968). F2-region electron concentration enhancement from inner radiation belt particles, *J. Geophys. Res.*, *73*(19). 6275-6283.

Krasovskii, V. I., et al. (1958). The observation of corpuscles by means of the third artificial earth satellite, (in Russian). *Iskusstvennye Sputniki Zemli*, *2*, 59-60 (English translation: *Planet. Space Sci.*, *5*, 248-249, 1961).

Krasovskii, V. I., et al. (1961). The detection of electrons with energies of approximately 10 keV in the upper atmosphere (in Russian). *Iskusstvennye Sputniki Zemli*, 6, 113-126, (English translation: *Planet. Space Sci.*, *9*, 27-40, 1962).

Krupnikov, K. K., Makletsov, A. A., Mileev, V. N., et al. (1999). Computer simulation of spacecraft/environment interaction, *Rad. Meas.*, *30*, 653-659.

Kudela, K., et al. (1992). Inner zone electron peaks observed by the ''Active" satellite, *J. Geophys. Res.*, *97*, 8681-8683.

Kudela, K. (2013). On cosmic rays and space weather in the vicinity of Earth, in *Homage to the Discovery of Cosmic Rays*, ed. by J.A. Perez-Peraza, 177-199, NOVA Sci., Inc., New York.

Lazutin, L. L. & Kozelova, T. V. (2012). Energetic electron injections to the inner magnetosphere during magnetic storms and magnetospheric substorms, in *Proc. XXXIV Annual Seminar "Physics of Auroral Phenomena"*, Apatity, Russia, 1-4 March.

Lazutin, L. & Kudela, K. (2012). The space object of magnetoplasma: Magnetosphere of Earth, in *Interstellar Medium: New Research*, Eds. B.M. Cancellieri, et al., 168-203, NOVA Sci., Inc., New York.

Lazutin, L. & Kuznetsov, S. (2008). Nature of sudden auroral activations at the beginning of magnetic storms, *Geomag. and Aeron.*, *48*, 2, 165-174.

Lemaire, J. F. (2003). The effect of a southward interplanetary magnetic field on Störmer's allowed regions, *Adv. Space Res.*, *31*, 1131-1153.

Li, X., Roth, I., Temerin, M., Wygant, J., Hudson, M. K. & Blake, J. B. (1993). Simulation of the prompt energization and transport of radiation particles during the March 23, 1991 SSC, *Geophys. Res. Lett.*, *20*(22). 2423–2426, doi:10.1029/93GL02701.

Lieu, R., et al. (1988). Observations of low-latitude electron precipitation, *J. Geophys. Res.*, *93*, 4131-4133.

Lin, C. S. & Yeh, H. C. (2005). Satellite observations of electric fields in the South Atlantic anomaly region during the July 2000 magnetic storm, *J. Geophys. Res.*, *110*, A03305, doi: 10.1029/2003JA010215.

Lin, R. P., Dennis, B. R., Hurford, G. J., et al. (2002). The Reuven Ramaty High-Energy Solar Spectroscopic Imager, (RHESSI)", *Solar Physics*, *210*, 3-32.

Lei, J., et al. (2008a). Observations and simulations of quasiperiodic ionospheric oscillations and large-scale traveling ionospheric disturbances during the December 2006 geomagnetic storm, *J. Geophys. Res.*, *113*, A06310, doi:10.1029/2008JA013090.

Lei, J., et al. (2008b). Observations and simulations of the ionospheric and thermospheric response to the December 2006 geomagnetic storm: Initial phase, *J. Geophys. Res.*, *113*, A01314, doi:10.1029/2007JA012807.

Luhmann, J. G. & Blake, J. B. (1978). Rocket altitude atmospheric X-rays from magnetospheric electrons at low and middle latitudes, *J. Atmos. Terr. Phys.*, *40*, 465-469.

Lum, K. S. K., Mohr, J. J., Barret, D., et al. (1997). Simulations and measurements of the background encountered by a high-altitude balloon-borne experiment for hard X-ray astronomy, *Nuclear Instruments and Methods in Physics Research, Section A*, *396*, 350-359.

Lyons, L. R. & Williams, D. J. (1984). *Quantitative Aspects of Magnetospheric Physics*, D. Reidel Pub. Co., Dordrecht Boston.

Lyons, L. R. (1995). A new theory for magnetospheric substorms, *J. Geophys. Res.*, *100*, A10, 19,069– 19,081.

Lyons, L. R., et al. (2009). Evidence that solar wind fluctuations substantially affect global convection and substorm occurrence, *J. Geophys. Res.*, *114*, A11306, doi:10.1029/ 2009JA014281.

Malaspina, D. M., Andersson, L., Ergun, R. E., Wygant, J. R., et al. (2014). Nonlinear electric field structures in the inner magnetosphere, *Geophys. Res. Lett.*, *41*, 5693–5701, doi:10.1002/2014GL061109.

Marov, M. Ya, Shematovich, V. I. & Bisicalo, D. V. (1996). Nonequilibrium aeronomic processes: A kinetic approach to the mathematical modeling, *Space Sci. Rev.*, *76*, 1-204.

Matsui, H., Torbert, R. B., Spence, H. E., Khotyaintsev, Yu. V. & Lindqvist, P.-A. (2013). Revision of empirical electric field modeling in the inner magnetosphere using Cluster data, *J. Geophys. Res. Space Physics*, *118*, 4119–4134, doi:10.1002/jgra.50373.

Mitsuda, K., Bautz, M., Inoie, H., et al. (2007). The X-ray observatory Suzaku, *Publ. Astron. Soc. Japan*, *59*, S1-S7.

Mendillo, M. (2006). Storms in the ionosphere: Patterns and processes for total electron content, *Rev. Geophys.*, *44*, RG4001, doi:10.1029/ 2005RG000193.

Nagai, T., Yukimatu, A. S., Matsuoka, A., Asai, K. T., Green, J. C., Onsager, T. G. & Singer, H. J. (2006). Timescales of relativistic electron enhancements in the slot region, *J. Geophys. Res.*, *111*, A11205, doi:10.1029/2006JA011837.

Neighbours, J. E. & Clark, G. W. (1974). A survey of trapped low energy electrons near the inner boundary of the inner radiation zone from the OSO-7, *Proceedings of the Workshop on Electron Contamination in X-Ray Astronomy Experiments*, edited by S.S. Holt, NASA GSFC Rep. X-661-74-130.

Nishimura, Y., et al. (2006). Storm-time electric field distribution in the inner magnetosphere, *Geophys. Res. Lett.*, *33*, L22102, doi: 10.1029/ 2006GL027510

Nishimura, Y., et al. (2009). Response of convection electric fields in the magnetosphere to IMF orientation change, *J. Geophys. Res.*, *114*, A09206, doi:10.1029/2009JA014277.

Novikov L. S., Mileev, V. N., Krupnikov, K. K., et al. (2008). Simultaneous investigation of magnetospheric plasma and space craft charging, *Adv. Space Res.*, *42*, 1307-1312.

Nusinov A. A., Avdyuschin, S. I., Ginzburg, E. A., et al. (2012). Efficiency of detecting penetrationg radiation in "Meteor" satellites (in russian). *Science and Education*, *8*, 197-210, doi: 10.7463/ 0812.0450381

Pedatella, N. M., et al. (2009). Observations of the ionospheric response to the 15 December 2006 geomagnetic storm: Long-duration positive storm effect, *J. Geophys. Res.*, *114*, A12313, doi:10.1029/2009JA014568.

Paulikas, G. A. (1975). Precipitation of particles at low and middle latitudes, *Rev. Geophys. Space Phys.*, *13*(5). 709-734.

Potemra, T. A. & Rosenberg, T. J. (1973). VLF propagation disturbances and electron precipitation at mid-latitudes, *J. Geophys. Res.*, *78*(10). 1572-1580.

Revnivtsev M., et al. (2012). MVN: x-ray monitor of the sky on Russian segment of ISS, in *Space Telescopes and Instrumentation 2012: Ultraviolet to Gamma Ray*, edited by T. Takahashi, S.S. Murray, J.-W. A. den Herder, Amsterdam, Netherlands, 1-6 July, 2012, *Proceedings of SPIE*, vol. *8443*, SPIE, Bellingham, WA, doi: 10.1117/12.925916.

Rodger, C. J., et al. (2010). Use of POES SEM-2 observations to examine radiation belt dynamics and energetic electron precipitation into the atmosphere, *J. Geophys. Res.*, *115*, A04202, doi:10.1029/2008JA014023.

Rosen, A. & Sanders, N. L. (1971). Loss and replenishment of electrons in the inner radiation zone during 1965–1967, *Journal of Geophysical Research*, 1971, *76*, 110-121

Sarcar, R., Mandal, S., Debnath, D., et al. (2011). Instruments of RT-2 experiment onboard CORONAS-PHOTON and their test and evaluation IV: background simulations using GEANT-4 toolkit, *Exp. Astron.*, *89*, 85-107, doi:10.1007/s10686-010-9208-z.

Savenko, I. A., et al. (1962). Soft particle radiation at an altitude of 320 km in the latitudes near the equator, (in russian). *Iskusstvennye Sputniki Zemli*, 13, 75-80, (English translation: *Planet. Space Sci.*, *11*, 431-436, 1963).

Seward, F. D., Grader, R. J., Toor, A., Burginyon, G. A. & Hill, R. W. (1974). Electrons at low altitudes: A difficult background problem for soft X-ray astronomy, *Proceedings of the Workshop on Electron Contamination in X-Ray Astronomy Experiments*, edited by S.S. Holt, NASA GSFC Rep. X-661-74-130.

Sheehan, C. H. & St.-Maurice, J.-P. (2004). Dissociative recombination of N2 +, O2 +, and NO+: Rate coefficients for ground state and vibrationally excited ions, *J. Geophys. Res.*, *109*, A03302, doi:10.1029/2003JA010132.

Shinbori, A., et al. (2006). Enhancements of magnetospheric convection electric field associated with sudden commencements in the inner magnetosphere and plasmasphere regions, *Adv. Space Res.*, *38*, 1595–1607, doi:10.1016/ j.asr.2005.05.082

Smith, R., Snowden, S., Kuntz, K. D. (Eds.), (2009). The Local Bubble and Beyond II, Philadelphia, USA, 21-24 April, 2008, *AIP Conference Proceedings/Astronomy and Astrophysics, 1156*.

Suparta, W. & Guzrizal. (2014). The variability of space radiation hazards towards LEO spacecraft, *J. Phys.: Conf. Ser.*, *539*, 012023, doi: 10.1088/1742-6596/539/1/012023

Suvorova, A. V., Tsai, L.-C. & Dmitriev, A. V. (2012a). On relation between mid-latitude ionospheric ionization and quasi-trapped energetic electrons during 15 December 2006 magnetic storm, *Planet. Space Sci.*, *60*, 363-369, doi:10.1016/j.pss.2011.11.001.

Suvorova, A. V., Tsai, L.-C. & Dmitriev, A. V. (2012b). On magnetospheric source for positive ionospheric storms, *Sun and Geosphere*, *7*, 2, 91-96.

Suvorova, A. V., Dmitriev, A. V., Tsai, L.-C., Kunityn, V. E., Andreeva, E. S., Nesterov, I. A. & Lazutin, L. L. (2013a). TEC evidence for near-equatorial energy deposition by 30 keV electrons in the topside ionosphere, *J. Geophys. Res.*, *118*, 4672–4695, doi:10.1002/jgra.50439.

Suvorova, A. V., Tsai, L.-C., Dmitriev, A. V. (2013b). TEC enhancement due to energetic Electrons above Taiwan and the West Pacific, *Terr. Atmos. Ocean. Sci.*, *24*, 2, 213-224, doi: 10.3319/TAO.2012.09.26.01

Suvorova, A.V., Dmitriev, A. V. & Tsai, L.-C. (2013c). Evidence for near-equatorial deposition by energetic electrons in the ionospheric F-layer, in *Proc. of International Scientific Online Conference on Modern Engineering and Technologies of the Future, (METF 2013).* ed. by A.Khnykin, Krasnoyarsk, Russia, 6-8 February 2013, 68-82.

Suvorova, A. V., Huang, C.-M., Matsumoto, H., Dmitriev, A. V., Kunityn, V. E., Andreeva, E. S., Nesterov, I. A. & Tsai, L.-C. (2014a). Low-latitude ionospheric effects of energetic electrons during a recurrent magnetic storm, *J. Geophys. Res. Space Physics*, *119*, 9283-9303, doi:10.1002/2014JA020349

Suvorova, A. V., Dmitriev, A. V. & Huang, C.-M. (2014b). Energetic electron enhancements below the radiation belt and X-ray contamination at low-orbiting satellites, *J. Astrophys.*, Article ID 701498, 5 pages, doi:10.1155/2014/701498.

Suvorova, A. V. & A. V. Dmitriev, (2014). On forbidden high-energy electrons as a source of background in X-ray and gamma-ray observations, *arXiv:1412.0482*

Suvorova, A. V., Huang, C.-M., Tsai, L.-C., Dmitriev, A. V. & Ratovsky, K. G. (2015). Long-duration ionospheric positive storm during the December 2006 geomagnetic storm: Ionizing effect of forbidden electrons, submitted to *Adv. Space Res.*, Jan 2015.

Takahashi, T., Murray, S. S. & den Herder, J.-W. A. (Eds.), (2012). *Space Telescopes and Instrumentation 2012: Ultraviolet to Gamma Ray*, Amsterdam, Netherlands, 1-6 July, 2012, *Proceedings of SPIE, 8443*, SPIE, Bellingham, WA, doi: 10.1117/12.2008904.

Winningham, J. D. (1974). Low energy, (10 eV to 10 keV) equatorial particle fluxes, *Proceedings of the Workshop on Electron Contamination in X-Ray Astronomy Experiments*, edited by S.S. Holt, NASA GSFC Rep. X-661-74-130.

Wygant, J. R., Mozer, F., Temerin, M., Blake, J., Maynard, N., Singer, H. & Smiddy, M. (1994). Large amplitude electric and magnetic field signatures in the inner magnetosphere during injection of 15 MeV electron drift echoes, *Geophys. Res. Lett.*, *21*, 1739–1742, doi:10.1029/94GL00375.

Wygant, J. R., Singer, H. J., Temerin, M., Mozer, F. & Hudson, M. K. (1998). Experimental evidence on the role of the large spatial scale electric field in creating the ring current, *J. Geophys. Res.*, *103*(A12). 29527–29544.

In: Cyclonic and Geomagnetic Storms
Editor: Victoria P. Banks
ISBN: 978-1-63482-360-9

Chapter 3

RECURRENT IONOSPHERIC STORMS DURING SOLAR MINIMUM

A. Dmitriev
Institute of Space Science, National Central University, Chung-Li, Taiwan,
Skobeltsyn Institute of Nuclear Physics Moscow State University, Moscow, Russia

ABSTRACT

We present a comprehensive analysis of perturbations in the low- and mid-latitude ionosphere during moderate recurrent geomagnetic storms (RGSs) in 2007 – 2008 solar minimum. It is shown that during so-called recurrent ionospheric storms (RIS), the heliospheric – magnetospheric – ionospheric - thermospheric coupling remains highly efficient even during solar minimum. Analysis of global ionospheric maps of vertical total electron content (VTEC) derived on the base of GPS network showed that VTEC has a tendency to decrease and negative ionospheric storms occur right before the onset and on the late recovery phase of RGS. A positive ionospheric storm is developed 2 to 4 days after, i.e., on the recovery phase of RGS produced by high-intensity long-duration continuous auroral activity. The ionospheric response to heliospheric and geomagnetic drivers reveals prominent longitudinal and local time variations. Such effects as penetrating electric fields of interplanetary and magnetospheric origin, disturbance dynamo electric fields, neutral winds and thermospheric chemical composition changes are considered as possible mechanisms responsible for the observed recurrent ionospheric storms.

Keywords: Ionospheric storms, recurrent geomagnetic storms, corotating interaction regions, high-speed solar wind

1. INTRODUCTION

The recent period of low solar activity was accompanied by prominent ionospheric variations related to so-called recurrent geomagnetic storms (RGSs) [Tsurutani et al., 1995; 2011; Verkhoglyadova et al.; 2011; Zhang and Paxton, 2011; Dmitriev et al., 2013]. RGSs and their ionospheric and thermospheric effects attract a great interest during recent years [e.g., Lei et al., 2008a;b;c; 2011; Mlynczak et al., 2008; Thayer et al., 2008; Denton et al., 2009; Sojka et al., 2009; Verkhoglyadova et al., 2011; Liu et al., 2012a]. Comprehensive studies of the ionospheric day-to-day variability reveal that geomagnetic and meteorological sources of the variations in the F2 layer ionization are comparable while the solar radiation source is a minor contributor [Forbes et al., 2000; Rishbeth and Mendillo, 2001]. The regularity of interplanetary driving and thermospheric/ionospheric responses suggests a possibility to develop a reliable 27-day prediction of RGS-related ionospheric disturbances, or so-called recurrent ionospheric storms (RIS).

Recurrent geomagnetic storms are generated in interaction of the magnetosphere with a complex interplanetary structure consisting of high-speed streams (HSSs) of the solar wind preceded by corotating interaction regions (CIR) formed between the fast and slow solar wind streams [Burlaga and Lepping, 1977; Tsurutani et al., 1995; 2011a]. During rising and declining phases and minimum of the solar cycle, RGSs occur almost every 7 to 13 days because of the draped heliospheric current sheet related to longitudinal distribution of active regions and coronal holes on the Sun. The most dramatic geomagnetic response to the HSSs are chains of consecutive substorms caused by the southward components of large-amplitude Alfven waves within the body of HSSs. This auroral activity has been named high-intensity long-duration continuous AE activity (HILDCAAs) [Tsurutani et al., 1995].

The RGSs are typically not very intense [Richardson et al., 2006]. They peak at *Dst* ~ -40 nT but have a prominent seasonal variation of ~40 nT which is ordered by the spring and fall equinoxes. The substorm activity is generally most intense near the peak of HSS where the Alfven wave amplitudes are greatest, and it decreases with decreasing wave amplitudes and solar wind speed. Hence, the RGSs have long-lasting main and recovery phases. The main phase is accompanied by relatively high solar wind dynamic pressure and rapidly varying strong IMF *Bz*, associated with the CIR. This phase is characterized by intense particle precipitation in the auroral region and from the radiation belt to lower latitudes. The recovery phase is accompanied by HSSs and related HILDCAAs, which contribute the fresh and sporadic injection of substorm energy leading to unusually long storm recovery phases (up to ~10 days) as noted in *Dst*. This phase is also often accompanied by relatively low solar wind density and weakened alternating IMF *Bz*.

Complex mechanism of the thermosphere-ionosphere system response to a geomagnetic storm involves a number of various agents such as disturbance electric fields, changes in neutral winds system and neutral chemical composition, gravity waves and diffusion [e.g., Mendillo, 2006]. Despite of relatively weakness, RGSs are accompanied by prominent enhancements (positive storm) and depletions (negative storms) of the ionospheric electron content [Tulasi Ram et al., 2010; Verkhoglyadova et al., 2011]. There are a number of differences between the RGSs and CME-driven geomagnetic storms and their ionospheric manifestations [Borovsky and Denton, 2006; Liu et al., 2012a]. The main difference consists in the penetration of the interplanetary electric field (IEF) in the magnetosphere. In contrast to prompt (time scale of minutes) penetration for CME-driven storms [e.g., Guo et al., 2011], during CIRs, the daytime ionospheric responses to IEF changes were found to occur within ~1 hour [Koga et al., 2011]. Burns et al. [2012] suggested that while CIRs are substantial factor affecting the thermosphere-ionosphere response, the level and duration of the continuing forcing during the HSSs is also important. This continuing forcing leads to extended periods of storm effects in the ionospheric electron content and enhanced neutral densities for many days after the end of the CIR phase. Thus, although the magnitudes of RGSs are considerably smaller than those of CME-driven storms, their cumulative effects on the thermosphere and ionosphere are comparable [e.g., Turner et al., 2009].

A distinct global response of the ionosphere to RGSs was reported in a number of papers [e.g., Denton et al., 2009; Ram et al., 2010; Pedatella et al., 2010; Pedatella and Forbes, 2011; Verkhoglyadova et al., 2011; Liu et al., 2012a]. On average, the electron content in the low-latitude ionosphere increases during the CIR-related main phase of RGSs. The virtual height of the F2 layer peak (hmF2) on the dayside is found to increase by a few tens of km. During the long-lasting recovery phase related to HSS, the electron content decreases but persists higher than the pre-storm level. Ram et al. [2010] pointed out that the RGSs are characterized by a strong altitude and latitude dependence of electron density perturbations, which can be explained by such physical processes as photoionization-chemistry, particle precipitation, and dynamic and diffusion transport. The disturbance dynamo electric field (DDEF) [Blanc and Richmond, 1980; Huang et al., 2005] is also considered to be an important mechanism for driving the electrodynamic response of the dayside ionosphere to recurrent geomagnetic activity related to high-speed solar wind streams [Pedatella et al., 2010; Pedatella and Forbes, 2011; Liu et al., 2012a].

Fast (within a few hours), global, and continuous ionospheric responses to RGSs were reported by a number of authors [e.g., Ram et al., 2010; Verkhoglyadova et al. 2011]. The largest ionospheric variations were found at low latitudes. It was concluded that CIRs/HSSs were external drivers for both thermospheric and ionospheric phenomena. A global thermospheric heating was revealed during RGSs. For instance, Sojka et al. [2009] found a significant increase of the ion temperature in the high-latitude ionosphere during CIR intervals. Prominent increases of low-latitude thermospheric density and ΣO/N2 during RGSs were also found in a number of studies [Lei et al., 2008a; 2011; Crowley et al., 2008; Burke et al., 2010; Liu et al., 2012b]. Burke et al. [2010] have shown that a model used to estimate exospheric temperature changes during large CME-driven storms overpredicts the thermospheric heating during RGSs. However, large-amplitude Alfvén waves in the interiors of HSSs generate regularly observed increases in the exospheric temperature. Deng et al. [2011] concluded that the energy transfer process into the upper atmosphere associated with HSSs is a combination of Joule heating and particle precipitation at high latitudes, while Joule heating plays a dominant role.

An important manifestation of efficient heliospheric – magnetospheric – ionospheric – thermospheric coupling occurring during RGSs is a 9-day periodicity in variations of the ionospheric and thermospheric parameters [Crowley et al., 2008; Lei et al., 2008a;b;c; Mlynczak et al., 2008; Thayer et al., 2008; Chang et al., 2009; Pedatella et al., 2010; Ram et al., 2010; Liu et al., 2010a; 2012a]. On average, at low-middle latitudes, the 9-day oscillations in thermospheric density, temperature and such ionospheric parameters as electron density in the maximum of F2 layer (NmF2), hmF2 and the thickness of F2 layer (HT) are in phase with those in geomagnetic activity.

However, in the equatorial anomaly region, the ionosphere shows more complicated day-to-day variability. Wang et al. [2011] found that low-latitude stations, located at different longitudes, detected ionospheric variations with different periodicities of 11 and 16 – 21 days, though the 9-day periodicity was dominant. The difference indicates a longitudinal inhomogeneity in the ionospheric day-to-day variability related to geomagnetic variations as well as to other sources. Dmitriev et al. [2013] have found a strong longitudinal inhomogeneity in the low-latitude ionosphere response to RGSs. Moreover, a negative ionospheric storm was revealed on the CIR-related initial phase of RGSs that might be caused by a strong heating of the low-latitude thermosphere.

At middle latitudes, Tsagouri et al. [2000] also found a strong dependence of negative storm effects occurrence on geomagnetic activity. These effects were observed as a response to geomagnetic storms and not during a substorm lifetime. Negative storm effects are triggered and driven by the solar wind energy deposition to the auroral ionosphere but they are furthermore modified by a number of contributing mechanisms. Prölss [1980] establishes that negative ionospheric storms are closely coupled to changes in the neutral composition. Model simulations [e.g., Fuller-Rowell et al., 1994; Burns et al., 2006] show that negative storms are caused by increases of molecular nitrogen in regions of sunlight, the strength of which depends on longitude and the local time of the sector during the storm input. Strickland et al. [2001] reported direct observations of the negative ionospheric storm occurred simultaneously with the molecular nitrogen enhancement. In addition, Chukwuma [2006] demonstrates an example of negative ionospheric storms produced by significant increase of the solar wind dynamic pressure.

In the present work, we report the results of comprehensive analysis of positive and negative ionospheric disturbances during recurrent magnetic storms observed in 2007 to 2008, i.e., in the minimum of solar activity.

2. Experimental Data

2.1. Solar and Geomagnetic Activity

Figure 1 shows that the solar minimum in 2007-2008 is characterized by low level of the solar radiation represented by the solar radio flux with wavelength of 10.7 cm (F10.7 index). The F10.7 index can be considered as a proxy of the solar radiation controlling the dayside ionospheric ionization [Ozguc et al., 2008; *A* et al., 2012].

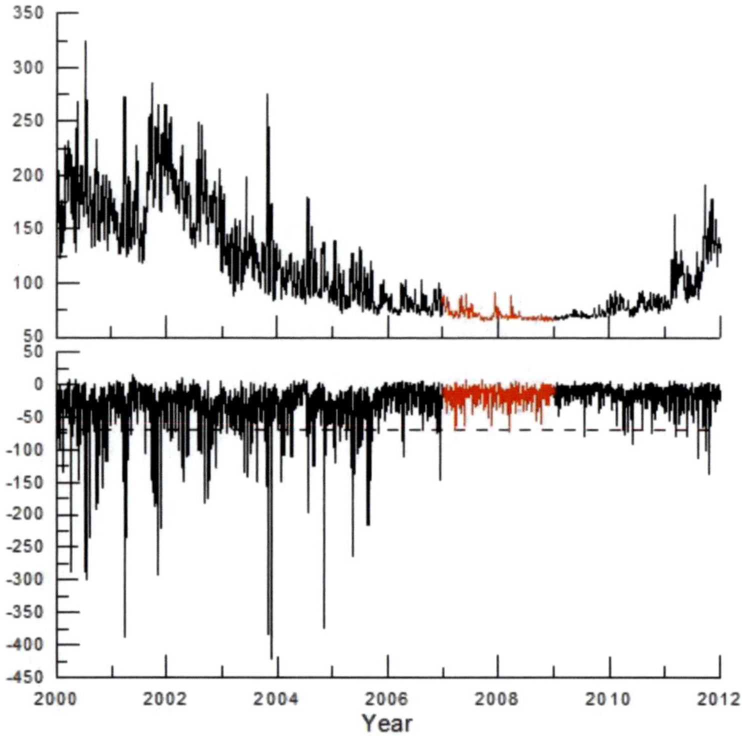

Figure 1. Solar cycle variations of solar radio flux F10.7 (top) and daily minimum Dst index (bottom) from 2000 to 2012. The interval used for the analysis from 2007 to 2008 (indicated by red) corresponds to solar minimum accompanied by corotating high-speed solar wind streams and recurrent geomagnetic storms (RGSs) of moderate and low intensity (Dstmin > -70 nT as indicated by the dashed line at the bottom).

During this time, the heliospheric and geomagnetic conditions are dominated, respectively, by CIR/HSSs and by RGSs, whose amplitude, measured as a minimum *Dst* variation, does not exceed 40 nT [e.g., Bothmer et al., 2004]. We do not consider the year of 2009 when a deep minimum occurred in solar, heliospheric, geomagnetic and ionospheric/thermospheric activity [Tsurutani et al., 2011b; Verkhoglyadova et al., 2013].

2.2. Global Ionospheric Maps

The long duration of RGSs make it possible to study ionospheric effects of these storms with 2-hour resolution. This time resolution is widely used for a global ionosphere mapping such 2-D global ionospheric maps (GIMs) of vertical total electron content (VTEC) derived from the global positioning system (GPS) network [Rebischung et al., 2012; http://aiuws.unibe.ch / ionosphere/]. GIMs are generated using data acquired from about 200 ground-based receivers of the radio-signals from GPS and GLONASS satellite constellations and analyzed by the International Global Navigation Satellite Systems Service (IGS) and other institutions [http://igscb.jpl.nasa.gov /network/refframe.html].

The reliability of GIMs was investigated in a number of studies [e.g., Hernandez-Pajares et al., 2009; Jee et al., 2010]. It was shown that on the whole, GIMs were largely able to reproduce the spatial and temporal variations of the global ionosphere as well as annual and semiannual variations and solar cycle variations. The map of receivers is shown in Figure 2. Above the Pacific Ocean (longitudes from 150° to 180° and from –180° to -120°), in particular its West-South part, the number of receivers is less than that above continents. However, each longitudinal range is represented by at least 4 receivers located in both Southern and Northern hemispheres at low- to mid-latitudes that allows for measuring a region of the equatorial ionization anomaly (EIA).

Figures 3 and 4 demonstrate examples of GIMs constructed during a recurrent magnetic storm on 1 – 8 January 2007 and compared with a quiet day on 13 January 2007. Solar, heliospheric and geomagnetic conditions during the storm are presented in Figure 5.

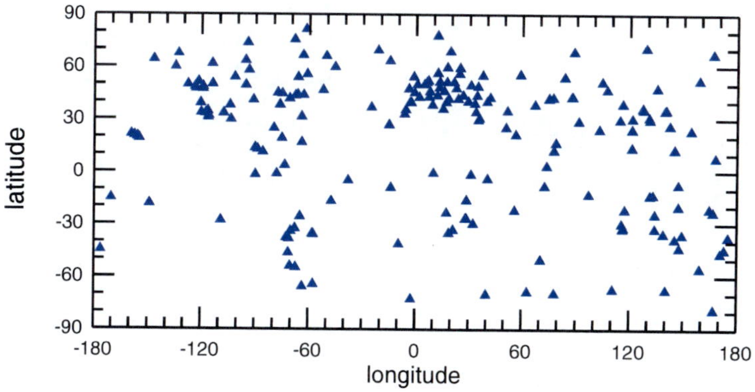

Figure 2. Geographic map of ground-based GPS receivers used for calculation of global ionospheric maps (GIMs) of vertical total electron content (VTEC). The receivers are distributed more or less uniformly, excepting the South-East Pacific region.

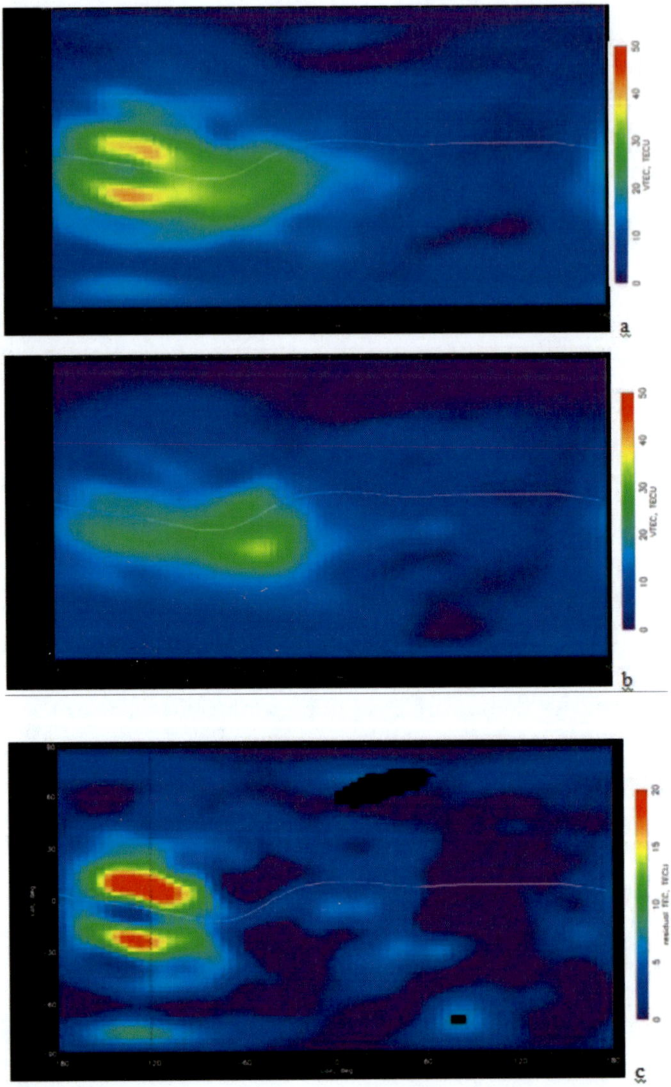

Figure 3. VTEC GIMs constructed at 20 UT during (a) disturbed day on 4 January 2007, (b) quiet day on 13 January 2007 and (c) residual between the disturbed and quiet days. The dip equator is shown by the white curve. The vertical black dashed line indicates local noon. Strong positive ionospheric storm occurred at low latitudes in the noon sectors in the Pacific region (at longitudes from -180° to –60°).

Namely, January 4 and 6 correspond to the recovery phase of the storm. In Figures 3, the residual VTEC reveals a prominent positive ionospheric storm, with enhancements up to >20 TECU occurring at low latitudes in the noon sector above Pacific region. In Figure 4, a prominent negative ionospheric storm with amplitude of >20 TECU can be revealed in the residual VTEC in the evening sector. It is important to note that the largest variations of VTEC occur in the crest of EIA.

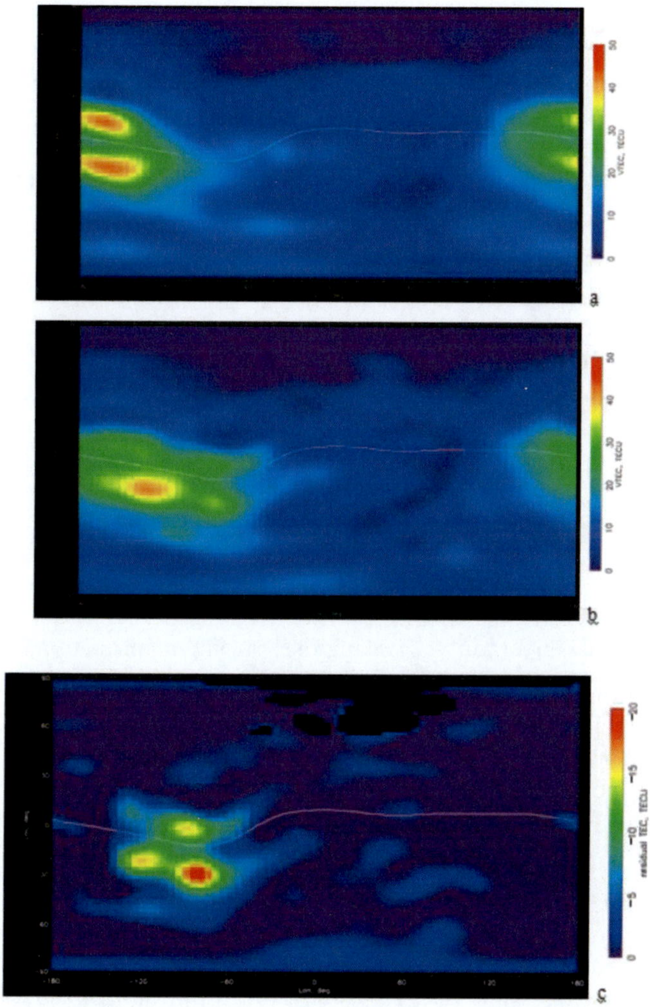

Figure 4. The same as in Figure 3 but for 0 UT at disturbed day on 6 January 2007 and quiet day on 14 January 2007 and (c) residual between the disturbed and quiet days. Strong negative ionospheric storm occurred at low latitudes in the evening sectors above the Pacific region (at longitudes from -120° to –40°).

2.3. Problem of Quiet Day

In order to calculate the residual VTEC, a quiet day has to be determined. This point is crucial in analysis of moderate or small ionospheric disturbances during RGSs when the variations of VTEC of ~20 TECU in the crests of EIA [see Dmitriev et al., 2013]. We search the quietest ionospheric state during the month to minimize statistical errors arising from a day-to-day variability of the ionosphere [Mendillo et al., 2010]. The errors of ~5 TECU in determination of the quiet or average VTEC value could influence our estimation of storm-time VTEC changes. In this connection, we reconsider a way of determining a quiet ionosphere.

Widely accepted is that a geomagnetically quiet period, without geomagnetic storm and substorm, corresponds to $Kp < 3$. It is assuming that during this period the ionosphere should be quiet. However, observations show that the ionosphere under $Kp = 1\text{-}3$ is still disturbed and deviations from monthly medians of ionospheric parameters can reach 20-40 % [Bilitza, 2001; Mendillo et al., 2010]. Actually, geomagnetic activity practically never ceases, resulting in the day-to-day ionospheric variability. Moreover, a period satisfying to the $Kp<3$ condition and determined as quiet can last many days, especially during solar minimum. In practice, researchers often used averaging data during the five quietest days of a month, monthly medians or a specific quiet day nearest to a storm day as well. However, from our point of view, the *Kp*-criterion used in many studies for determining the ionospheric ground state seems to be very rough.

To minimize uncertainty, we impose stronger restrictions on driving factors. A set of criteria is applied to the solar wind plasma and interplanetary magnetic field (IMF), solar and geomagnetic indices. Particularly, we use $Kp<1$ and $AE<100$ nT conditions for geomagnetic activity. Also additional constraints imposed on fast and large variations of IMF and the high-speed solar wind that can cause a moderate or weak auroral activity [Tsurutani et al., 1995]. Note that slow solar wind streams statistically dominate during solar minimum [Dmitriev et al., 2000] and always lead to a decay of geomagnetic activity. A separation between fast and slow streams coincides with an average solar wind speed of ~400 km/s [Dmitriev et al., 2000; 2009; Veselovsky et al., 2010]. Moreover, the value of 400 km/s is also a threshold for developing the Kelvin-Helmholtz instability on the magnetospheric flanks, which also intensifies the auroral activity [e.g., Singer et al., 1977; Lee and Olson, 1980]. Therefore, it is important to select time periods when the solar wind velocity is less than 400 km/s.

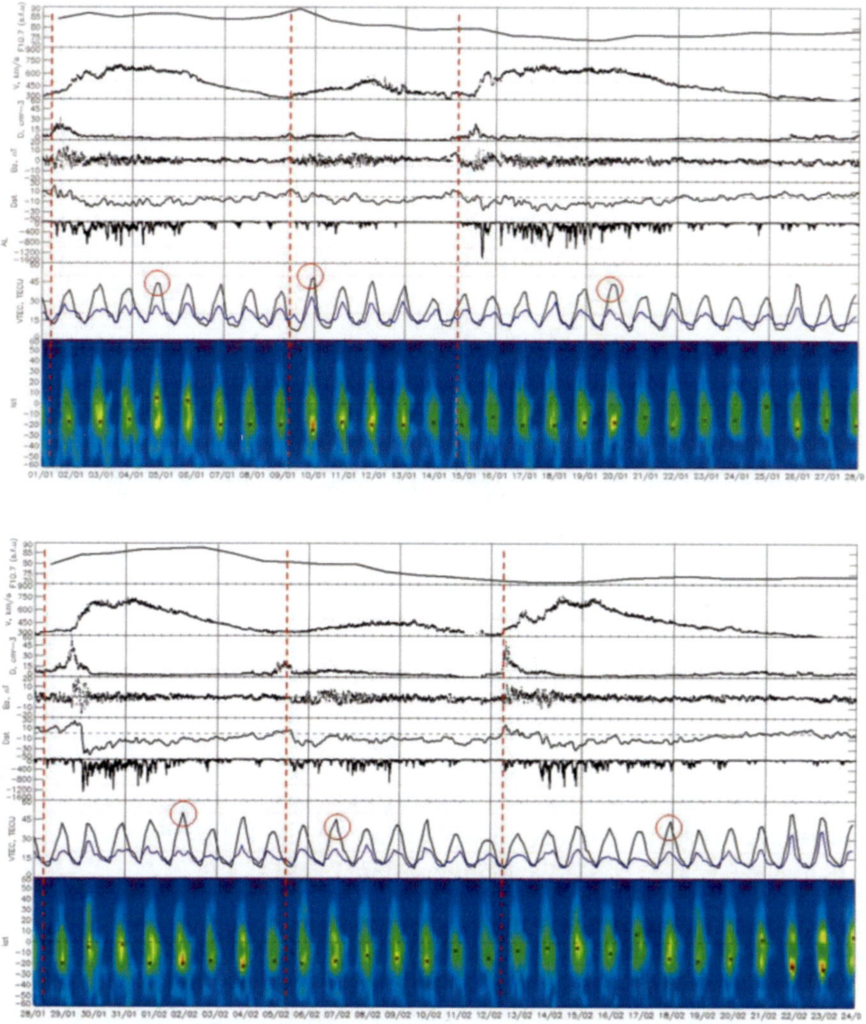

a

b

Figure 5. 27-day intervals of solar, heliospheric, geomagnetic and ionospheric variations from 1 to 27 January 2007 (a) and from 28 January to 23 February 2007. Panels show (from top to bottom): solar radio flux F10.7, solar wind velocity, density, IMF Bz component in GSM, geomagnetic indices Dst, AL, variations of maximal VTEC at low (black curve) and middle (blue curve) latitudes and variations of VTEC (colored 2D histogram) in longitudinal ranges from -120° to -90°. The black stars indicate maxima of VTEC. The red vertical dashed lines indicate the onset of RGSs. The red circles depict peaks of VTEC. VTEC increases during RGSs.

Table 1. List of quiet days in 2007 – 2008

Year	Month	Day		Year	Month	Day
2007	1	13		2008	4	3
2007	3	19		2008	4	14
2007	4	14		2008	5	15
2007	5	12		2008	6	13
2007	6	6		2008	7	8
2007	7	9		2008	7	20
2007	7	25		2008	8	2
2007	8	5		2008	8	29
2007	9	13		2008	9	13
2007	10	11		2008	9	27
2007	11	8		2008	10	9
2007	12	8		2008	10	27
2008	1	3		2008	11	5
2008	1	30		2008	11	22
2008	2	26		2008	12	2
2008	3	25		2008	12	15

Actually, only one or two days during each ~27 day interval are characterized by the lowest solar wind speed, and therefore to the quietest state of the ionosphere-magnetosphere system. Based on this fact in Figure 5, we can choose consistent quiet days of 13 January and 21 February 2007. The list of all quiet days in 2007 – 2008 is presented in Table 1.

3. Recurrent Storms

3.1. Recurrent Variations of VTEC

The 2-hour time step of GIMs is equivalent to a 30° step in longitude. That allows for dividing the map onto 12 longitudinal ranges: from -180° to -150°, from -150° to -120°, from 150° to 180°. We can, thus, construct a portion of GIM every two hours and for each longitudinal range. Spatial VTEC distribution varies with local time and, hence, with UT such that higher (lower) VTEC appear on the dayside (nightside).

Figure 5 demonstrates variations of VTEC distributions within two consequent 27-day intervals from 1 to 27 January and from 28 January to 23 February 2007 in the longitudinal range from –120° to –90°. The interval of 27 days is approximately equal to a period of solar rotation in the Earth's frame that is a prime source of the recurrent variations. We also show time profiles of daily solar radiation (F10.7), and 1-min data of solar wind velocity (*Vsw*), density (*D*), IMF *B*z in GSM coordinates, and 1-hour geomagnetic indices *AL* and *Dst*. Solar wind parameters were acquired from ACE and Wind upstream monitors. The Wind monitor was used when the ACE data was poor due to numerous data gaps. For the days when neither ACE nor Wind data were available, we eliminated the solar wind data from consideration.

During the both 27-day periods, one can identify 3 intervals of co-rotating CIR-HSS structures, characterized by high solar wind speeds of >400 km/s. The total duration of these structures varies from 4 to 10 days. The co-rotating streams result in recurrent geomagnetic storms of similar duration and repeating time. The onset of RGS is determined as an abrupt increase of geomagnetic indices *AL* and *Dst*. The increase results from magnification of southward IMF in compressed solar wind of the CIR region. The time interval between the onsets varies from 7 to 13 days. The storms are characterized by a high level of the auroral activity (large *AL* index) as well as by global geomagnetic disturbances revealed in enhanced *Dst* index (with minimum *Dst* of about –40 nT).

As one can clearly see in Figure 5, VTEC also enhanced during recurrent magnetic storms that occurred on 1 – 7 January (8-day duration), 9 - 12 January (4-day duration), 14 - 25 January (12-day duration), 29 January – 4 February (7-day duration), 5 – 10 February (6-day duration) and 12 – 21 February (12-day duration). At low latitudes (from –30° to 30°), VTEC usually peaks several days after the onset of RGSs, namely, on January 4, 10, 19 and February 1, 7 and 17. At middle latitudes (from –60° to -30° and from 30° to 60°), the situation is not so unambiguous: VTEC can peak in the first two days of RGSs as well as several days after the onset. It is interesting to note that the CIR-related onsets of RGS on January 2, 14, 29, February 5 and 12 were accompanied by relatively small VTEC at low latitudes. At that time, the IMF *B*z was often negative and large. During the late recovery phase of RGSs, VTEC has a tendency to decrease. The decrease was not gradual but sustained a day-to-day variability with local minima and maxima occurred at different longitudes during different days.

Hence, VTEC peaks usually 2 to 4 days after the RGS onset and sometimes even on the late stage of the recovery phase of RGSs. Most prominent enhancements of VTEC were rather related to HSSs. The amplitude of VTEC variations during RGSs can be estimated as being in the range of tens of TECU.

It is important to point out that the ionosphere and geomagnetic activity is disturbed by RGSs during most of time. There are only a few quiet days in the ionosphere per 27-day solar rotation period. Those days are characterized by both low geomagnetic activity and quiet solar wind. Similar patterns can be found for all other 27-day intervals in 2007 – 2008. Therefore, RGS-related ionospheric effects contribute a significant portion of the statistics.

3.2. Residual VTEC

In contrast to positive ionospheric storms, negative ones are much more difficult to investigate. Quantitative analysis of the positive and negative ionospheric storms is based on residual VTEC obtained from subtraction of a quiet VTEC from a storm-time VTEC. Figure 6 shows GIMs of residual VTEC constructed at 20 UT (left) and 24 UT (right) during RGS on 1 – 7 January 2007. The quiet day was selected on 13 January 2007 as we discussed in the Section 2.3.

As one can see in Figure 6, strong positive and negative ionospheric of ~20 TECU magnitude occurred at low latitudes over Eastern Pacific in the noon and evening sectors, respectively. At middle latitudes, the ionospheric storms were weaker and they occupied whole postnoon and evening sectors and some portion of them occurred even in the morning sector.

The temporal dynamics of positive and negative ionospheric storms was different. A strong positive ionospheric storm was developed at low and middle latitudes on January 4 and 5, i.e., on the 4th and 5th day of the RGS. Strong negative ionospheric storms occurred at low latitudes on 1 January and from 4 to 7 January, i.e., right in the beginning and on the recovery phase of the RGS. At middle latitudes, relatively intense (~5 TECU) negative ionospheric storms were developed on 1 January in the morning sector and on the late recovery phase from 6 to 7 January in the evening and morning sectors.

Similar analysis of residual VTEC GIMs was conducted for 70 RGSs, which we have identified in 2007 - 2008. The list of RGSs analyzed is presented in Table 2. For all 2-hour intervals of UT, positive and negative ionospheric storms were identified at various longitudes and local time.

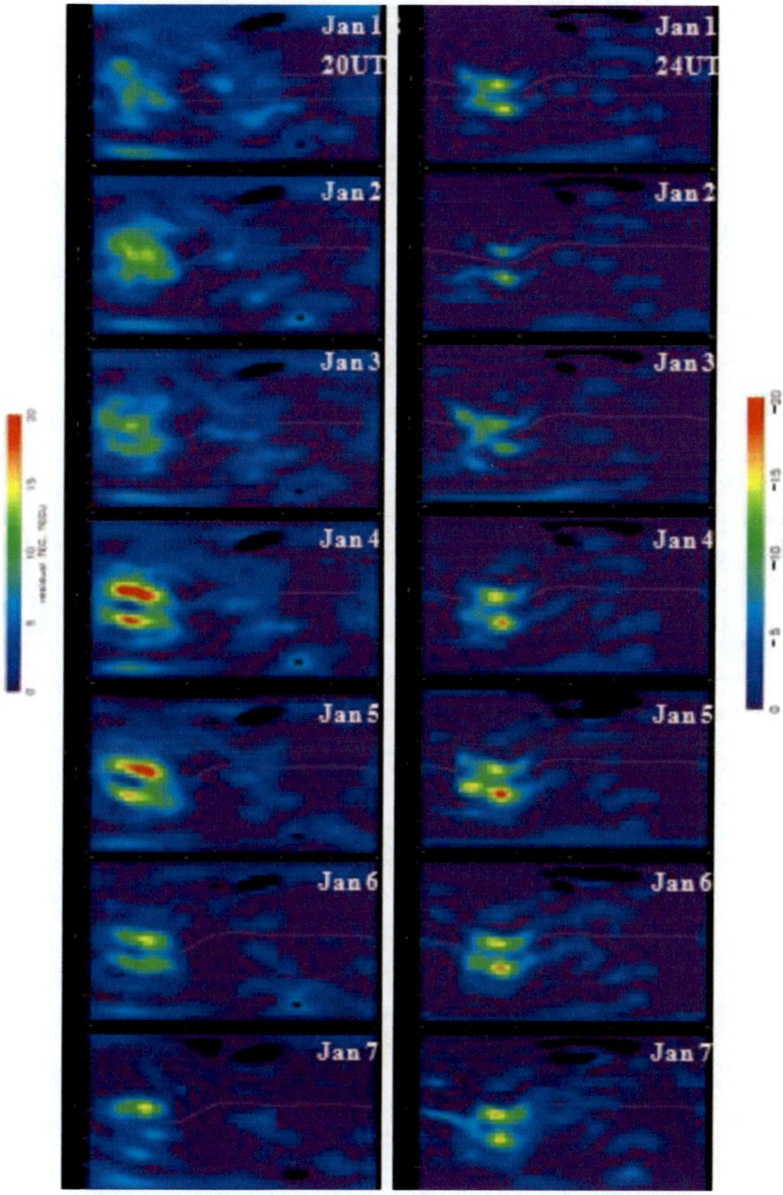

Figure 6. Residual VTEC GIMs constructed at 20 UT (left) and 24 UT (right) during RGS on January 2007. The dip equator is shown by the white curve. Vertical black dashed lines indicate local noon. Strong positive (left) and negative (right) ionospheric storms occur over Pacific at low latitudes in the noon and evening sectors, respectively.

Table 2. List of recurrent geomagnetic storms (RGSs)

N	Year	Month	Day	UT	Duration		N	Year	Month	Day	UT	Duration
1	2007	1	1	6	7		36	2007	11	12	12	7
2	2007	1	9	6	5		37	2007	11	19	18	11
3	2007	1	14	12	12		38	2007	12	9	12	7
4	2007	1	29	0	6		39	2007	12	16	12	9
5	2007	2	12	6	11		40	2008	1	5	0	7
6	2007	2	25	18	7		41	2008	1	12	12	11
7	2007	3	5	6	5		42	2008	1	24	12	6
8	2007	3	11	6	10		43	2008	1	31	12	7
9	2007	3	23	0	8		44	2008	2	10	0	8
10	2007	3	31	12	8		45	2008	2	18	6	9
11	2007	4	8	18	7		46	2008	2	27	12	9
12	2007	4	22	0	5		47	2008	3	8	6	14
13	2007	4	26	18	8		48	2008	3	26	0	7
14	2007	5	7	0	7		49	2008	4	3	0	11
15	2007	5	17	0	4		50	2008	4	15	12	6
16	2007	5	22	6	9		51	2008	4	22	12	7
17	2007	6	1	0	6		52	2008	4	30	12	11
18	2007	6	13	6	6		53	2008	5	19	0	7
19	2007	6	21	6	6		54	2008	5	27	12	9
20	2007	6	28	12	4		55	2008	6	14	12	9
21	2007	7	3	6	6		56	2008	6	24	18	9
22	2007	7	10	18	3		57	2008	7	10	6	9
23	2007	7	14	6	4		58	2008	7	21	0	9
24	2007	7	20	6	5		59	2008	8	8	18	7
25	2007	7	26	12	10		60	2008	8	17	0	9
26	2007	8	6	6	4		61	2008	9	3	0	9
27	2007	8	10	12	4		62	2008	9	14	18	10
28	2007	8	14	18	7		63	2008	9	30	12	9
29	2007	8	25	0	6		64	2008	10	10	12	7
30	2007	8	31	12	12		65	2008	10	26	0	10
31	2007	9	20	6	7		66	2008	11	7	0	7
32	2007	9	27	12	5		67	2008	11	14	12	7
33	2007	10	2	18	6		68	2008	11	24	18	7
34	2007	10	17	12	7		69	2008	12	3	0	7
35	2007	10	25	12	12		70	2008	12	22	12	7

The onset was determined with accuracy of 3 hour in four local time sectors: midnight (LT = 21 – 3), morning (LT = 3 – 9), noon (LT = 9 – 15), and evening (LT = 15 – 21). Dmitriev et al. [2013] showed that the occurrence probability of onsets does not depend on the local time. Hence, longitudinal differences in the ionospheric response (if any) to RGSs should weakly depend on the differences in local time of onsets.

3.3. Superposed Epoch Analysis

In order to study numerically the ionospheric recurrent storms, we use a superposed epoch analysis of 70 RGSs occurred from 2007 to 2008. The list of storms is presented in Table 2. The time is calculated from the onset of RMS (day = 0). The negative and positive time corresponds to periods, respectively, before and after the storm onset. The superposed epoch analysis is performed for daily values. Note that for each storm, the time starts (day = 0) from the onset such that the analyzed days are different from the calendar days. Within 10-day time intervals starting 2 days before the onset and lasting 8 days after (if possible), we calculate for each day a minimum of *Dst, Kp, AL* and *Bz*, maximum of *Vsw*, *D*, and *AE* as well as maximal amplitudes of positive and negative residual VTEC in various longitudinal sectors. Note that RGSs with duration less than 8 days contribute to the statistics of the first few days only.

Figure 7 demonstrates average variations of solar wind and geomagnetic parameters during 70 RGSs in 2007-2008. Before RGS (day = -2 and –1), the solar wind velocity and density are low and *Bz* ~ 0 that promote a quiet geomagnetic activity. The beginning of RGS (day = 0) is caused by a CIR region, in which the solar wind density and negative *B*z reach maximum values that result in intensification of geomagnetic activity and growth of *Kp* and *AE* indices. On the second day (day = 1), *B*z persists negative and high while the solar wind velocity increases but the density decreases that results in generation of storm maximum with peak values of the geomagnetic indices. On average, *Kp* and *AE* demonstrate quite well anti-correlation with the *Dst* index. It is interesting to note that the solar wind velocity peaks on the third day of RGS (day = 2), i.e., on the next day after the RGS maximum. At that time, the southward IMF decreases slightly, the solar wind density approaches to its nominal value of ~5 cm^{-3} and the geomagnetic activity is diminishing.

It is important to note that during RGSs, day-to-day geomagnetic variability in *Dst*–index is produced by two competitive drivers: solar wind pressure *P*d and IMF *Bz*. Positive variations in *Dst* are produced by northward turning IMF (*Bz* > 0) as well as by enhancements in the *P*d (solar wind density and/or velocity) resulting in magnetospheric compression and intensification of Chapman-Ferraro current at the magnetopause [e.g., O'Brien and McPherron, 2002]. Increases of southward IMF (*Bz* < 0) lead to negative *Dst* variations. It seems that the small negative *Dst* ~ -10 nT during the RGS onset results from high *P*d whose geomagnetic effect is stronger than the effect of southward IMF. The latter can be strong but its duration is usually short that does not result in large negative *Dst* variation.

Figure 8 shows variations of daily amplitudes of positive and negative ionospheric disturbances at low latitudes (from -30° to 30°) averaged over 70 RGSs in four longitudinal ranges: all longitudes, from 120° to 150°, from 0° to 30°, and from -120° to -90°. These ranges are well covered by IGS receivers as one can see in Figure 2. The amplitudes are set to be equal to 0 at the day = -2. The residual VTEC during positive storms has a tendency to decrease and forms a negative storm right before the onset and in the late recovery phase of RGSs. The wide maximum of positive ionospheric storms can be found from the first (day = 0) to the third (day = 2) day of RGS. On average, the residual VTEC variation correlates better with the solar wind velocity than with the geomagnetic indices. Such dynamics of VTEC allow us to introduce so-called recurrent ionospheric storms related rather to CIR/HSS solar wind structures than to RGS alone.

In different longitudinal ranges, the ionospheric storms have different behavior. The most intense positive storms are revealed above American region in the longitude range from –120° to –90° that is in good agreement with the results reported by Dmitriev et al. [2013]. The negative storms occur on the day = -1, i.e., right before the storm onset, and on the late recovery phase. The prolonged negative storms begin on average from the 4-th day of RGS above Pacific and Asia and from the 6-th day Above Africa.

Figure 9 shows variations of daily amplitudes of positive and negative ionospheric disturbances at middle latitudes from –60° to -30° and from 30° to 60° averaged over 70 RGSs in four longitudinal ranges. As in Figure 8, the amplitudes are set to be equal to 0 at the day = -2. During tow days before the storm (day = 1 and 2), the residual VTEC is close to 0 indicating no ionospheric storm. A strong positive ionospheric storm peaks on the first day and second days of RGS (day = 0 and 2). On the recovery phase the positive storm is gradually diminishing and then transfers into a negative storm on the late recovery phase (day = 5 and later).

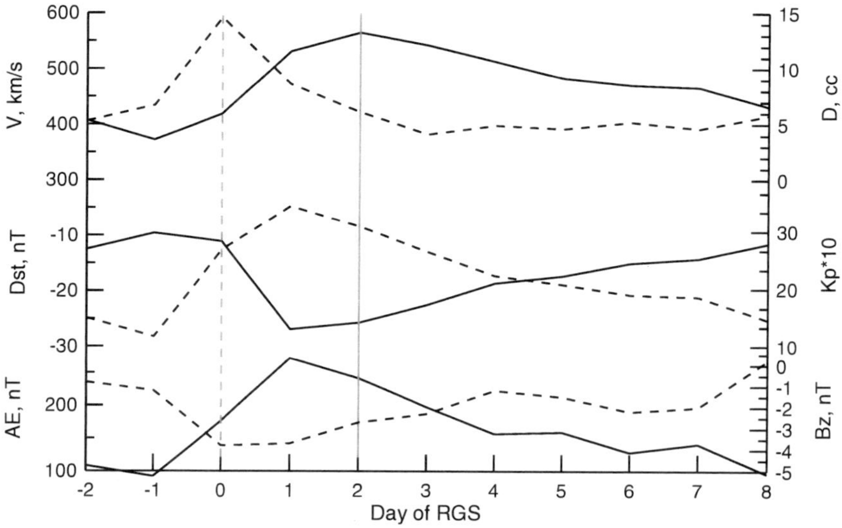

Figure 7. Average variations of various parameters during RGSs (from top to bottom): maximal solar wind velocity (solid curve, left axis) and density (dashed curve, right axis); minimal Dst (solid curve, left axis) and maximal Kp (dashed curve, right axis); maximal AE (solid curve, left axis) and minimal Bz (dashed curve, right axis). The onset of RMS corresponds to the day = 0 (vertical dashed line). Geomagnetic indices peak one day after the onset. Maximum of the solar wind velocity occur 2 days after the onset (indicated by vertical solid line).

The dynamics of positive ionospheric storm anticorrelates with the *Dst* variations rather than with the solar wind velocity. Note that the pattern of mid-latitude positive storms varies substantially with longitudes. The most intense and prolonged positive storm is developed at longitudes from 0° to 30°. In contrast, the longitudinal range from 120° to 150° is characterized by a minor and short-duration positive storm. Over the American sector (longitudes from -120° to -90°), the positive storm has a moderate intensity and duration.

The dynamics of negative storm at middle latitudes is substantially different from that at low latitudes. As one can see in Figure 9, a mid-latitude negative storm is developed on recovery phase of RGS. The prominent long-lasting negative storms are developed from the 2nd day of RGS (day = 1) over the Asian and 0-meridian sectors. Over the American sector, a negative storm is developed from the 4th day of RGS.

Simultaneous occurrence of both positive and negative ionospheric storms in the same latitudinal and longitudinal ranges indicates to their difference in local time as we demonstrated in Figure 6.

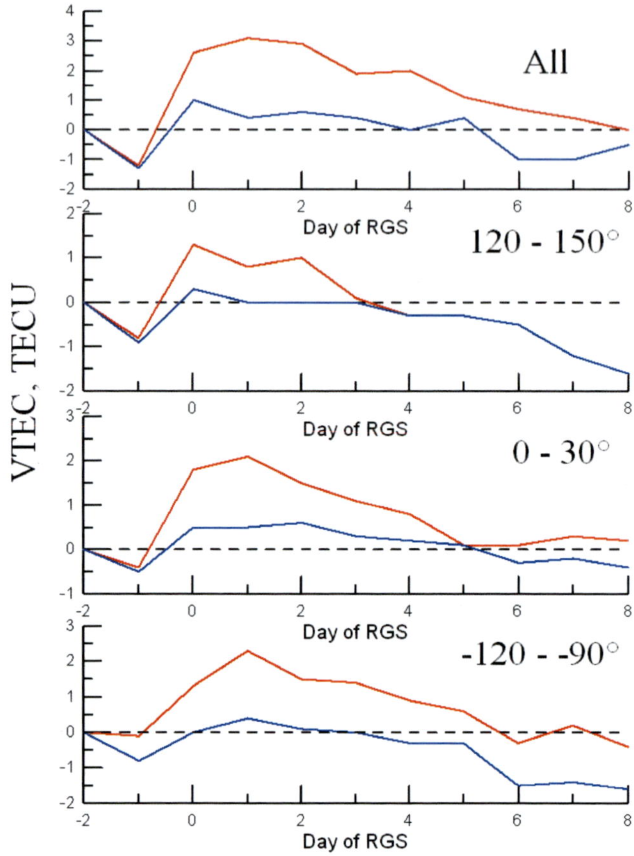

Figure 8. Average variations of amplitudes of positive (red curves) and negative (blue curves) ionospheric storms at low latitudes (from -30° to 30°) during RGSs in various longitudinal ranges (from top to bottom): all longitudes, from 120° to 150°, from 0° to 30°, and from -120° to -90°. The amplitudes are set to be equal to 0 at the day = -2.

Figure 10 shows a statistical distribution of local time of peak residual VTEC for positive ionospheric storms at low (from -30° to 30°) and middle (from –60° to -30° and from 30° to 60°) latitudes. The statistics was accumulated form 2-hour residual GIMs in 2007-2008. It can be clearly seen that the positive storms occur mostly in the noon and postnoon sectors from 11 to 16 LT. In the evening sector, the probability of positive storm occurrence is still high. There are practically no positive storms from the premidnight to the morning sector.

The LT distribution of negative storm is totally different from the negative ones as can be seen in Figure 11. Vast majority of the negative storms both at low and middle latitudes occur in the postnoon and evening sectors from 14 to 19 LT. Low-latitude negative storms do not occur from the premidnight to the noon sectors. However, at middle latitudes, the negative storms can occur quite often in the morning sector from 4 to 9 LT.

The difference in local time distribution between the positive and negative storms indicates to a difference in the mechanisms of their generation.

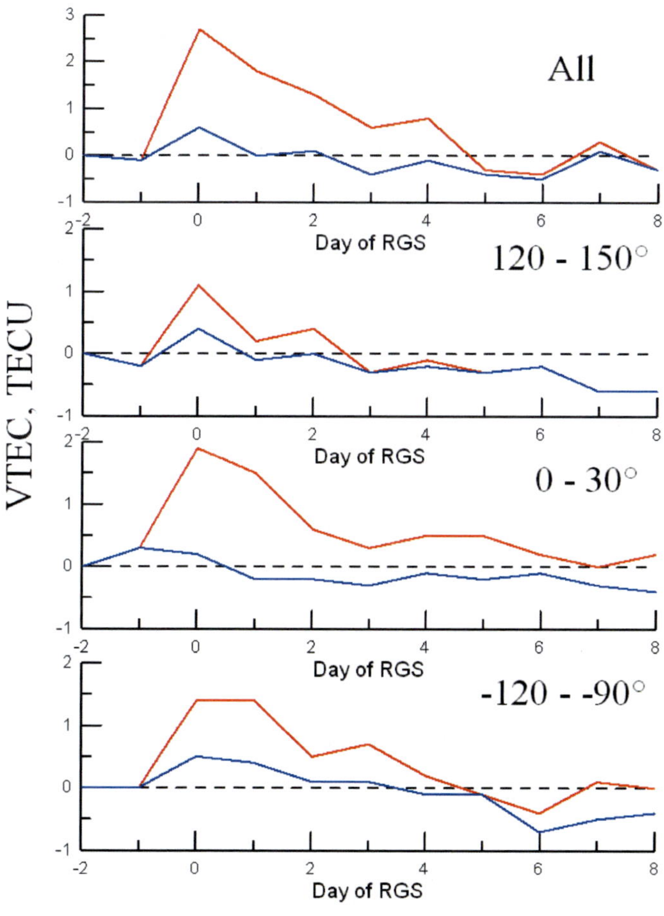

Figure 9. The same as in Figure 8 but for middle latitudes from –60° to -30° and from 30° to 60°.

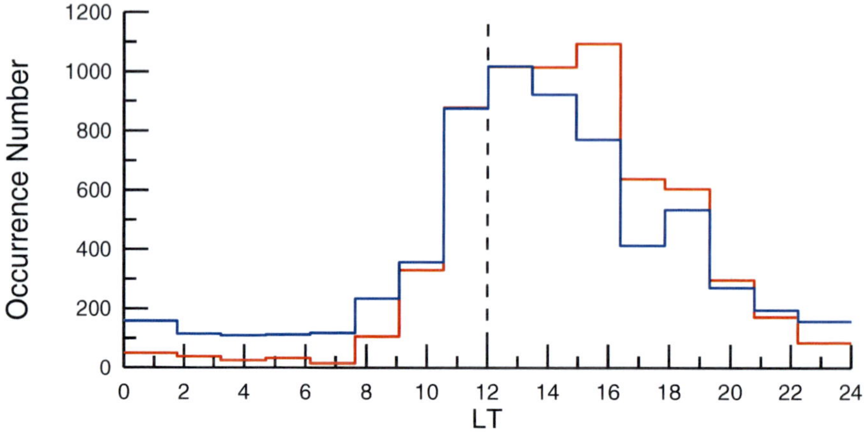

Figure 10. Statistical distribution of local time (LT) of peak values of positive ionospheric storms at low (from -30° to 30°) and middle (from –60° to -30° and from 30° to 60°) latitudes. The positive storms occur mostly in the noon and postnoon sectors from 11 to 16 LT.

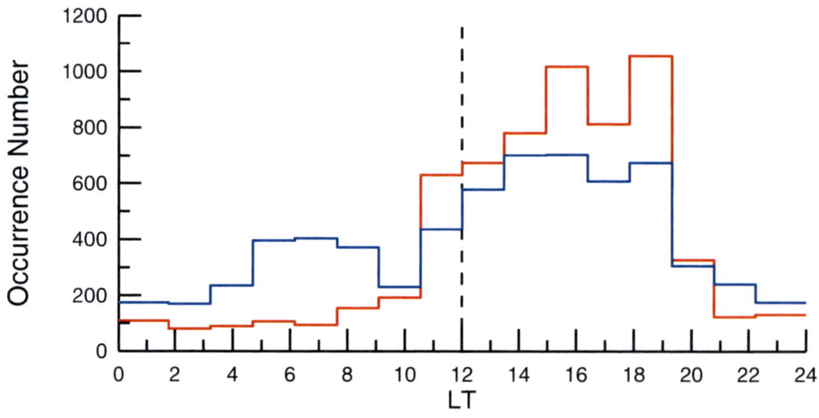

Figure 11 The same as in Figure 10 but for negative ionospheric storms. The negative storms occur mostly in the postnoon and evening sectors at LT from 14 to 19. There is also a secondary maximum in the morning sector for mid-latitude negative storms.

4. DISCUSSION AND SUMMARY

We have conducted a comprehensive statistical analysis of global ionospheric maps during solar minimum in 2007 – 2008. It was found that recurrent geomagnetic storms are accompanied by recurrent positive and negative ionospheric storms at low and middle latitudes. The storms are revealed as prominent enhancements and decreases of residual VTEC. We have also found that the amplitude of recurrent ionospheric storms varies substantially with longitude. The longitudinal variations in the RGS-driven ionosphere are difficult to explain by the effects typical for CME-driven storms such as an effect of "storm start time" [Fuller-Rowell et al., 1994].

As reported by Dmitriev et al. [2013], the RGSs onsets are practically equally probable at different UT. Moreover, RGSs last for several days and largest variations of VTEC occur several days after the onset and storm maximum (see Figures 8 and 9) such that the ionosphere losses information not only about the onset but also about the main phase of a storm. In contrast to CME-driven storms, an RGS is generated by two geoeffective solar wind structures: CIR and HSS. The onset and main phase of RGS are related to CIR. The prolonged maximum and recovery phase are related to HSS.

In the beginning of RGS, VTEC does not exhibit any prominent enhancement and even has a tendency to be suppressed at low latitudes such that a negative low-latitude ionospheric storm occur on the day = -1 (see Figures 5, 6, 8 and 9). The suppression of VTEC can be related to a strong magnetospheric compression caused by high solar wind dynamic pressure in the CIR region as reported by Chukwuma [2006].

The suppression of VTEC in the crest of EIA in the beginning of RGS can be also caused by an increase of recombination rate. Verkhoglyadova et al. [2011] reported an excess of thermospheric heating at low to high latitudes during CIR intervals. This excess was interpreted as additional energy input from geomagnetic activity heating. The heating causes a global increase in neutral density, which exceeds 40% at 400 km altitude in the low-latitude dayside ionosphere [Lei et al., 2011; Liu et al., 2012b]. The increase of neutral density does not affect the vertical integral of TEC. However, the increase of temperature results in an increase of the recombination rate and, thus, in a decrease of the ionization.

On the other hand, a continuous increase of $\Sigma O/N_2$ at low latitudes during RGSs was reported [*Crowley et al.*, 2008; Liu et al., 2012b]. The change of chemical composition of neutral species is attributed to dawnward vertical winds, which carry atomic-oxygen-rich air to lower latitudes that result in enhancements in F-region electron densities. The vertical winds are primarily driven by meridional neutral winds blowing toward equator from the hot auroral regions where Joule and particle heating is produced by continuous geomagnetic activity.

Hence in the beginning of recurrent storms, the VTEC dynamics are affected by two opposite factors: atmospheric heating, which increases the recombination rate, and downward vertical winds enriched by atomic oxygen, which increase the electron density. In addition, VTEC enhancements, observed sometimes during CIR intervals, can be produced by strong PPEF of interplanetary origin when IMF turns southward.

The CIR interval is followed by HSS. During HSS, VTEC increases substantially in the maximum of RGSs and during the early recovery phase of RGSs on the second - third day after the onset (see Figures 8 and 9). From analysis of EC vertical profiles, Dmitriev et al. [2013] showed that the maximum of RGS-related positive ionospheric storm is characterized by elevations of hmF2 by a few tens of kilometers and thickening of the F2 layer, while the latitude of crests in EIA remains practically unchangeable. This pattern is very close to that revealed by Ram et al. [2010] from COSMIC/FORMOSAT-3 RO measurements of the ionospheric disturbances during RGSs. It was suggested that the observed changes in EC could be related to enhancements in the thermospheric temperature and neutral composition $\Sigma O/N_2$.

In the low-latitude thermosphere, HSS intervals are accompanied by a decrease of the density and temperature [Verkhoglyadova et al., 2011, Liu et al. 2012b]. Hence, higher peak and scale heights of the F2 layer in the maximum of the positive ionospheric storm cannot be explained by the thermospheric factors. However, an excess of the ratio $\Sigma O/N_2$ at low latitudes still persists [Crowley et al., 2008; Liu et al. 2012b]. The excess of $\Sigma O/N_2$ results from continuous heating at middle and high latitudes produced by geomagnetic activity [Sojka et al., 2009; Verkhoglyadova et al., 2011], which is generated by high-amplitude IMF variations inherent to HSS. The high ratio $\Sigma O/N_2$ can explain an increase of VTEC at low and middle latitudes but it is unable to explain the changes of peak and scale heights.

The elevation of nmF2 and thickening of the F2 layer as well as an increase of VTEC might be also related to intensification of the fountain effect. Note that the effect of DDEF in the noon, postnoon and evening sectors results in suppression or even reversal of the eastward equatorial electrojet and, thus, this mechanism causes a suppression of the fountain effect [Huang et al., 2005; Huang, 2012] and development of a negative ionospheric storm at low latitudes. On the other hand, we find that the negative ionospheric storm occurs at low-latitude on the late recovery phase, when the operation of DDEF is diminishing. Hence, the mechanism of DDEF cannot help us to describe the recurrent ionospheric storms at low and middle latitudes.

The fountain effect on the dayside could be intensified by a prompt penetrating IEF (PPEF) of eastward direction. Studying GIMs with 2-hour resolution does not allow for detecting directly the effect of PPEF. During RGS, the effect is of short duration because of quickly varying IMF. In Figure 7, one can see that the daily maximum values of negative Bz remain strong within three days after the RGS onset. Hence, the observed dayside enhancements of 2-hour averaged VTEC might manifest in the operation of eastward PPEF somewhere within the 2-hour interval.

We have found that the maximum of low-latitude positive ionospheric storm occurs both in the maximum of RGS and on the day of the maximum solar wind velocity (see Figure 8) on the recovery phase of RGS. The long-lasting recovery phase of RGSs is produced by HILDCAAs, which in turn are related to HSSs characterized by large variations of IEF. Ultra-low-frequency fluctuations of IEF have a substantial effect on global convection and are an important contributor to the large-scale transfer of solar wind energy to the magnetosphere-ionosphere system [Lyons et al., 2009]. Hence, the fountain effect might be intensified by penetrating electric fields of the magnetospheric origin associated with the enhanced convection during HSSs.

A strong longitudinal variation was found in the recurrent ionospheric storms. In our analysis, the ionospheric storms are characterized by daily extrema of residual VTEC, which occur at roughly the same local times (noon and postnoon sectors for the positive storms and postnoon-evening sectors for negative ones). Hence, we are implicitly sampling the ionosphere in a constant local time reference frame. In such a constant local time frame, various nonmigrating tidal components will alias into stationary planetary wave components [e.g., Rishbeth and Mendillo, 2001; Forbes et al., 2006; Liu et al., 2010b]. However, the tide-related waves are mainly attributed to magnetic quiet periods with $Kp < 3$ and the amplitude of these waves is relatively small [Kil et al., 2012]. The lack of tidal effect might be also explained by the geomagnetic nature of the VTEC variations during RGSs. Namely, driving parameters, responsible for the geomagnetic disturbances, cannot capture the wave perturbations produced by lower atmospheric sources.

Kil et al. [2012] ruled out the possibility that the tidal modulation of the dynamo electric fields in the ionosphere is the source of longitudinal variations. They suggested mechanisms of the solar zenith angle and magnetic declination, which controls neutral winds and neutral composition at ionospheric heights [Rishbeth, 1998]. The effect of solar zenith angle is originated from a difference in the illumination of a dip equator, which governs the EIA region. Because the geomagnetic dipole is tilted and shifted relative to the axis of Earth's rotation, the dip coordinates do not coincide with the geographic ones and, thus, the ionospheric dynamics suffers from the effects of magnetic dip and declination [e.g., Challinor and Eccles, 1971].

Numerous statistical studies of ionospheric day-to day variability do not take into account the recurrent geomagnetic activity or used a criterion of $Kp < 3$. As one can see in Figure 8, the maximum of positive ionospheric storms corresponds to the average $Kp \sim 3$. Hence, it is important to estimate the contribution of RGSs to the statistics of ionospheric disturbances. For this purpose, we calculate an integral probability of occurrence of solar wind streams with various velocities from 250 to 800 km/s for the time interval from 2004 to 2008 (see Figure 12). Note that the average solar wind speed is ~400 km/s [Dmitriev et al., 2009; Veselovsky et al., 2010]. HILDCAAs are related to solar streams with speed >400 km/s [Tsurutani et al., 1995]. In Figure 5, one can clearly see that positive ionospheric storms persist during whole intervals of HILDCAAs and even 1 to 2 days longer. Hence, we can estimate approximately that the occurrence of positive ionospheric storms corresponds to the occurrence of >400 km/s solar wind streams. From Figure 12, we find that the probability of such occurrence is 0.6.

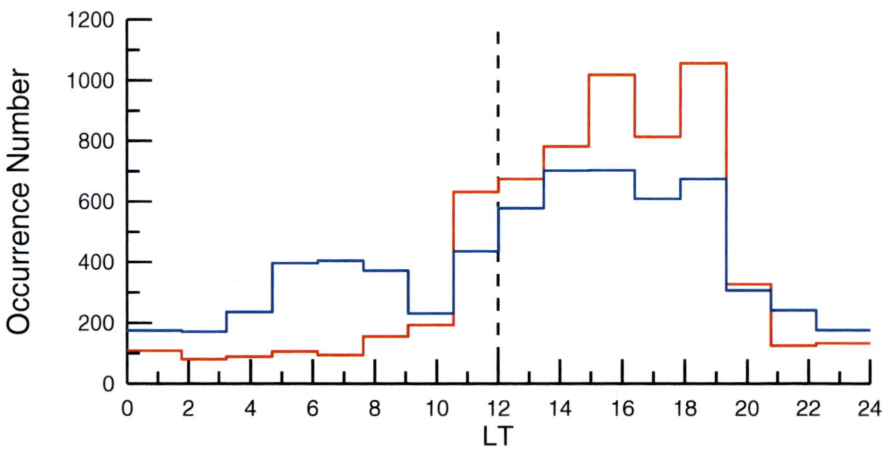

Figure 11 The same as in Figure 10 but for negative ionospheric storms. The negative storms occur mostly in the postnoon and evening sectors at LT from 14 to 19. There is also a secondary maximum in the morning sector for mid-latitude negative storms.

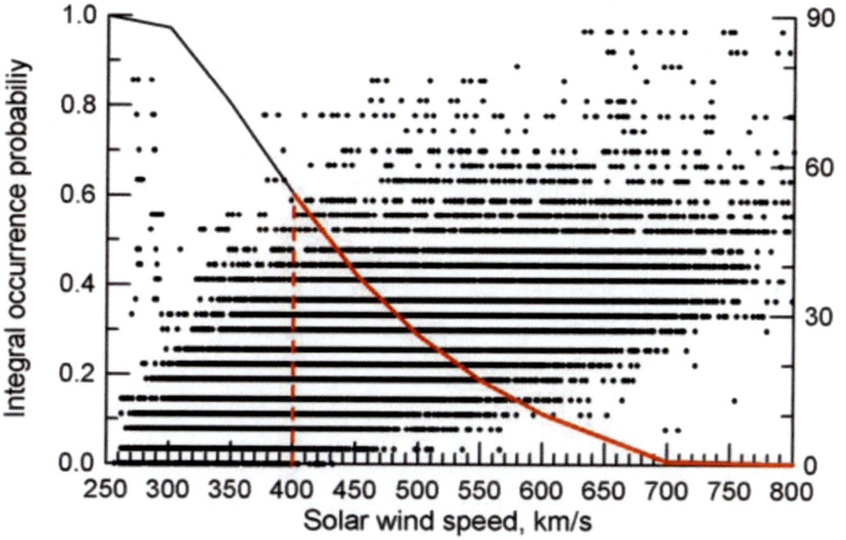

Figure 12. Integral probabilities: (solid curve, left axis) of occurrence of solar wind streams with velocities higher than given on declining phase and solar minimum in 2004 – 2008. Dots depict the scatter plot of *Kp* versus speed. High-speed solar wind streams with velocity >400 km/s (depicted by red curve) occur in 60% of statistics. More than 30% of statistics is characterized by small Kp < 3 under high-speed solar wind streams that is proper for RGSs.

A scatter plot of 3-hour *K*p index versus solar wind speed *V*sw is also shown in Figure 12. On average, *K*p increases with *V*sw. However, the spread is very wide. There are numerous events of small *K*p accompanied by high solar wind speed. We have estimated that 33% (39%) of statistics is characterized by $Kp < 3$ ($Kp \leq 3$) under $Vsw > 400$ km/s. Such conditions are proper for RGSs and related ionospheric disturbances. As a result, more than >30% of "quiet" ionospheric conditions determined from Kp < 3 can be actually contributed by ionospheric storms originated from HSSs.

Summarizing we can suggest that the recurrent ionospheric storms result from a complex of competitive phenomena such as thermospheric heating, change of neutral composition $\Sigma O/N_2$, PPEF, penetrating electric fields of the magnetospheric origin related to HILDCAAs, effect of solar zenith angle and probably magnetic declination effect. At low solar activity, including solar minimum, CIR/HSS result in ionospheric disturbances lasting up to 60% of time or even more. Hence, the recurrent ionospheric storms dominate in the statistics of ionospheric conditions.

ACKNOWLEDGMENTS

The authors thank Kyoto World Data Center for Geomagnetism (http://swdcwww.kugi.kyoto-u.ac.jp/index.html) for providing the *Dst*, *Kp*, *AE* and *AL* geomagnetic indices. The Global Ionosphere Maps/VTEC Data are produced by European Data Center (http://www.aiub.unibe.ch/ionosphere/). The ACE solar wind data were provided by N. Ness and D.J. McComas through the CDAWeb website. This work was supported by grants MOST-103-2111-M-008-016- from the Ministry of Science and Technology of Taiwan, by the Project-Oriented Research Program of National Central University of Taiwan and partially by grant NSC103-2923-M-006-002-MY3/14-05-92002HHC_a of Taiwan-Russia Research Cooperation.

REFERENCES

A, E., et al. (2012), A global model: Empirical orthogonal function analysis of total electron content 1999-2009 data, J. Geophys. Res., 117, A03328, doi:10.1029/2011JA017238.

Bilitza, D. (2001), International Reference Ionosphere 2000, *Radio Sci.*, *36*(2), 261, doi:10.1029/2000RS002432.

Blanc, M. and A.D. Richmond (1980), The ionospheric disturbance dynamo, *J. Geophys. Res.*, 85(A4), 1669-1686, doi:10.1029/JA085iA04p01669.

Borovsky, J. E., and M. H. Denton (2006), Differences between CME-driven storms and CIR-driven storms, *J. Geophys. Res.*, 111, A07S08, doi:10.1029/2005JA011447.

Bothmer V., et al. (2004), The solar and interplanetary causes of space storms in solar cycle 23, *IEEE Transactions on Plasma Science*, 32(4), Part 1, 1411 - 1414.

Burke, W. J., L. C. Gentile, and M. P. Hagan (2010), Thermospheric heating by high-speed streams in the solar wind, *J. Geophys. Res.*, 115, A06318, doi:10.1029/2009JA014585.

Burlaga, L.F., and R. P. Lepping (1977), The causes of recurrent geomagnetic storms, *Planet. Space Sci.*, 25, 1151.

Burns, A. G., W. Wang, T. L. Killeen, S. C. Solomon, and M. Wiltberger (2006), Vertical variations in the N2 mass mixing ratio during a thermospheric storm that have been simulated using a coupled magnetosphere-ionosphere-thermosphere model, *J. Geophys. Res.*, 111, A11309, doi:10.1029/2006JA011746.

Burns, A. G., et al. (2012), The effects of corotating interaction region/high speed stream storms on the thermosphere and ionosphere during the last solar minimum, *J. Atmos. Solar-Terr. Phys.*, 83, 79-87, doi:10.1016/j.jastp.2012.02.006.

Challinor, R. A. and D. Eccles (1971), Longitudinal variations of the mid-latitude ionosphere produced by neutral-air winds-I. Neutral-air winds and ionospheric drifts in the northern and southern hemispheres, *J. Atmos. Solar-Terr. Phys.*, 33(3), 363-369.

Chang, L. C., et al. (2009), Isolation of the global MLT thermal response to recurrent geomagnetic activity, *Geophys. Res. Lett.*, 36, L15813, doi:10.1029/2009GL039305.

Chukwuma, V. U. (2006), On the geomagnetic and ionospheric responses to an intense storm associated with weak IMF Bz and high solar wind dynamic pressure, *Acta Geophysica*, 55(4), 469-489, DOI: 10.2478/s11600-007-0030-6

Crowley, G., A. Reynolds, J. P. Thayer, J. Lei, L. J. Paxton, A. B. Christensen, Y. Zhang, R. R. Meier, and D. J. Strickland (2008), Periodic modulations in thermospheric composition by solar wind high speed streams, *Geophys. Res. Lett.*, 35, L21106, doi:10.1029/2008GL035745.

Deng, Y., et al. (2011), Energy input into the upper atmosphere associated with high-speed solar wind streams in 2005, *J. Geophys. Res.*, 116, A05303, doi:10.1029/2010JA016201.

Denton, M. H., T. Ulich, and E. Turunen (2009), Modification of midlatitude ionospheric parameters in the F2 layer by persistent high-speed solar wind streams, *Space Weather*, 7, S04006, doi:10.1029/2008SW000443.

Dmitriev, A.V., A.V. Suvorova, I.S. Veselovsky (2009), Statistical Characteristics of the Heliospheric Plasma and Magnetic Field at the Earth's Orbit during Four Solar Cycles 20-23, in Handbook on Solar Wind: Effects, Dynamics and Interactions, Ed. Hans E. Johannson, NOVA Science Publishers, Inc., New York, 2009, p. 81-144.

Dmitriev, A. V., A. V. Suvorova, and I. S. Veselovsky (2000), Solar wind and interplanetary magnetic field parameters at the Earth's orbit during three solar cycles, *Phys. Chem. Earth(C)*, *25*(1-2), 125-128.

Dmitriev, A. V., C.-M. Huang, P. S. Brahmanandam, L. C. Chang, K.-T. Chen, L.-C. Tsai, (2013), Longitudinal variations of positive dayside ionospheric storms related to recurrent geomagnetic storms, *J. Geophys. Res.*, *118*, doi:10.1002/jgra.50575.

Forbes, J. M., et al. (2006), Troposphere-thermosphere tidal coupling as measured by the SABER instrument on TIMED during July-September 2002, *J. Geophys. Res.*, 111, A10S06, doi:10.1029/2005JA011492.

Fuller-Rowell, T. J., M. V. Codrescu, R. J. Moffett, and S. Quegan (1994), Response of the Thermosphere and Ionosphere to Geomagnetic Storms, *J. Geophys. Res.*, 99(A3), 3893–3914, doi:10.1029/93JA02015.

Guo J., et al. (2011), Interplanetary drivers of ionospheric prompt penetration electric fields, *J. Atmos. Solar-Terr. Phys.*, 73, 130–136.

Hernandez-Pajares, M., et al. (2009), The IGS VTEC maps: A reliable source of ionospheric information since 1998, *J. Geod.*, 83, 263–275, doi:10.1007/s00190-008-0266-1.

Huang, C.-S. (2012), Equatorial ionospheric electrodynamics associated with high-speed solar wind streams during January-April 2007, *J. Geophys. Res.*, 117, A10311, doi:10.1029/2012JA017930.

Huang, C.-M., A. D. Richmond, and M.-Q. Chen (2005), Theoretical effects of geomagnetic activity on low-latitude ionospheric electric fields, *J. Geophys. Res.*, 110, A05312, doi:10.1029/2004JA010994.

Jee, G., et al. (2010), Assessment of GPS global ionosphere maps (GIM) by comparison between CODE GIM and TOPEX/Jason TEC data: Ionospheric perspective, *J. Geophys. Res.*, 115, A10319, doi:10.1029/2010JA015432.

Kil, H., et al. (2012), Persistent longitudinal features in the low-latitude ionosphere, *J. Geophys. Res.*, 117, A06315, doi:10.1029/2012JA017570.

Koga D., et al. (2011), Electrodynamic coupling processes between the magnetosphere and the equatorial ionosphere duringa 5-day HILDCAA event, *J. Atmos. Solar-Terr. Phys.*, 73, 148–155.

Lee, L. C., and J. V. Olson (1980), Kelvin-Helmholtz instability and the variation of geomagnetic pulsation activity, *Geophys. Res. Lett.*, *7*(10), 777-780, doi:10.1029/GL007i010p00777.

Lei, J., et al. (2008a), Rotating solar coronal holes and periodic modulation of the upper atmosphere, *Geophys. Res. Lett.*, 35, L10109, doi:10.1029/2008GL033875.

Lei, J., et al. (2008b), Ionosphere response to solar wind high-speed streams, *Geophys. Res. Lett.*, 35, L19105, doi:10.1029/2008GL035208.

Lei, J., et al. (2008c), Global thermospheric density variations caused by high-speed solar wind streams during the declining phase of solar cycle 23, *J. Geophys. Res.*, 113, A11303, doi:10.1029/2008JA013433.

Lei, J., et al. (2011), Impact of CIR storms on thermosphere density variability during the solar minimum of 2008, *Sol. Phys.*, 274, 427-437, doi:10.1007/s11207-010-9563-y.

Liu, J., et al. (2010a), Response of the topside ionosphere to recurrent geomagnetic activity, *J. Geophys. Res.*, 115, A12327, doi:10.1029/2010JA015810.

Liu, H., S. V. Thampi, and M. Yamamoto (2010b), Phase reversal of the diurnal cycle in the midlatitude ionosphere, *J. Geophys. Res.*, 115, A01305, doi:10.1029/2009JA014689.

Liu, J., et al. (2012a), High-speed stream impacts on the equatorial ionization anomaly region during the deep solar minimum year 2008, *J. Geophys. Res.*, 117, A10304, doi:10.1029/2012JA018015.

Liu, J., et al. (2012b), Superposed epoch analyses of thermospheric response to CIRs: Solar cycle and seasonal dependencies, *J. Geophys. Res.*, 117, A00L10, doi:10.1029/2011JA017315

Lyons, L. R., et al. (2009), Evidence that solar wind fluctuations substantially affect global convection and substorm occurrence, *J. Geophys. Res.*, 114, A11306, doi:10.1029/2009JA014281.

Mendillo, M. (2006), Storms in the ionosphere: Patterns and processes for total electron content. *Reviews of Geophysics*, 44, RG4001.doi:10.1029 /2005RG000193.

Mendillo, M., J. Wroten, R. Roble, and H. Rishbeth (2010), The equatorial and low-latitude ionosphere within the context of global modeling, *J. Atmos. Sol. Terr. Phys.*, *72*, 358-368, doi:10.1016/j.jastp.2009.03.026.

Mlynczak, M. G., et al. (2008), Solar-terrestrial coupling evidenced by periodic behavior in geomagnetic indexes and the infrared energy budget of the thermosphere, *Geophys. Res. Lett.*, 35, L05808, doi:10.1029/2007GL032620.

O'Brien, T. P., and R. L. McPherron (2002), Seasonal and diurnal variation of Dst dynamics, *J. Geophys. Res.*, 107(A11), 1341, doi:10.1029/2002JA 009435.

Ozguc, A., T. Atac, and R. Pektas (2008), Examination of the solar cycle variation of foF2 for cycles 22 and 23, *J. Atmos. Solar-Terr. Phys.*, 70, 268-27.

Pedatella, N. M., and J. M. Forbes (2011), Electrodynamic response of the ionosphere to high-speed solar wind streams, *J. Geophys. Res.*, 116, A12310, doi:10.1029/2011JA017050.

Pedatella, N. M., J. Lei, J. P. Thayer, and J. M. Forbes (2010), Ionosphere response to recurrent geomagnetic activity: Local time dependency, *J. Geophys. Res.*, 115, A02301, doi:10.1029/2009JA014712.

Prölss, G. W. (1980), Magnetic storm associated perturbations of the upper atmosphere: Recent results obtained by satellite-borne gas analyzers, *Rev. Geophys.*, 18(1), 183–202, doi:10.1029/RG018i001p00183.

Ram, Tulasi S., J. Lei, S.-Y. Su, C. H. Liu, C. H. Lin, and W. S. Chen (2010), Dayside ionospheric response to recurrent geomagnetic activity during the extreme solar minimum of 2008, *Geophys. Res. Lett.*, 37, L02101, doi:10.1029/2009GL041038.

Rebischung, P., et al. (2012), IGS08: the IGS realization of ITRF2008, *GPS Solut.*, 16, 483-494, DOI 10.1007/s10291-011-0248-2.

Richardson, I. G., et al. (2006), Major geomagnetic storms (Dst <-100 nT) generated by corotating interaction regions, *J. Geophys. Res.*, 111, A07S09, doi:10.1029/2005JA011476.

Rishbeth, H., (1998), How the thermospheric circulation affects the ionospheric F2-layer, *J. Atmos. Solar-Terr. Phys.*, 60, 1385 - 1402.

Rishbeth, H., and M. Mendillo (2001), Patterns of F2-layer variability, *J. Atmos. Solar-Terr. Phys.*, 63, 1661-1680.

Singer H. J., C. T. Russell, M. G. Kivelson, E. W. Greenstadt, and J. V. Olson (1977), Evidence for the control of Pc 3,4 magnetic pulsations by the solar wind velocity, *Geophys. Res. Lett.*, *4*(9), 377–379, doi: 10.1029/GL004i0 09p00377.

Sojka, J. J., R. L. McPherron, A. P. van Eyken, M. J. Nicolls, C. J. Heinselman, and J. D. Kelly (2009), Observations of ionospheric heating during the passage of solar coronal hole fast streams, *Geophys. Res. Lett.*, 36, L19105, doi:10.1029/2009GL039064.

Strickland, D. J., R. E. Daniell, and J. D. Craven (2001), Negative ionospheric storm coincident with DE 1-observed thermospheric disturbance on October 14, 1981, *J. Geophys. Res.*, 106, 21,049–21,062, doi:10.1029 /2000JA000209.

Thayer, J. P., et al. (2008), Thermospheric density oscillations due to periodic solar wind high-speed streams, *J. Geophys. Res.*, 113, A06307, doi:10.1029/2008JA013190.

Tsagouri, I., A. Belehaki, G. Moraitis, and H. Mavromichalaki (2000), Positive and negative ionospheric disturbances at middle latitudes during geomagnetic storms, *Geophys. Res. Lett.*, 27(21), 3579-3582, doi:10.1029/2000GL003743.

Tsurutani, B. T., W. Gonzalez, A. Gonzalez, F. Tang, J. Arballo, and M. Okada (1995), Interplanetary origin of geomagnetic activity in the declining phase of the solar cycle, *J. Geophys. Res.*, 100(A11), 21717-21733.

Tsurutani B. T., E. Echer, F. L. Guarnieri, W. D. Gonzalez (2011a), The properties of two solar wind high speed streams and related geomagnetic activity during the declining phase of solar cycle 23, *J. Atmos. Solar-Terr. Phys.*, 73 164–177.

Tsurutani, B. T., Echer, E., and Gonzalez, W. D. (2011b), The solar and interplanetary causes of the recent minimum in geomagnetic activity (MGA23): a combination of midlatitude small coronal holes, low IMF BZ variances, low solar wind speeds and low solar magnetic fields, *Ann. Geophys.*, 29, 839-849, doi:10.5194/angeo-29-839-2011

Turner, N. E., et al. (2009), Geoefficiency and energy partitioning in CIR-driven and CME-driven storms, *J. Atmos. Solar-Terr. Phys.*, 71, 1023-1031.

Verkhoglyadova, O. P., B. T. Tsurutani, A. J. Mannucci, M. G. Mlynczak, L. A. Hunt, A. Komjathy, and T. Runge (2011), Ionospheric VTEC and thermospheric infrared emission dynamics during corotating interaction region and high-speed stream intervals at solar minimum: 25 March to 26 April 2008, *J. Geophys. Res.*, 116, A09325, doi:10.1029/2011JA016604.

Verkhoglyadova, O. P., et al. (2013), Variability of ionospheric TEC during solar and geomagnetic minima (2008 and 2009): external high speed stream drivers, *Ann. Geophys.*, 31, 263-276, doi:10.5194/angeo-31-263-2013.

Veselovsky, I. S., A. V. Dmitriev, and A. V. Suvorova (2010), Algebra and statistics of the solar wind, *Cos. Res., 48*(2), 113-128, doi:10.1134/S0010952510020012.

Wang, W., et al. (2011), Ionospheric day-to-day variability around the whole heliosphere interval in 2008, *Sol. Phys.*, 274, 457-472, doi:10.1007/s11207-011-9747-0.

Zhang, Y., and L. J. Paxton (2011), Long-term variation in the thermosphere: TIMED/GUVI observations, *J. Geophys. Res.*, 116, A00H02, doi:10.1029/2010JA016337.

INDEX

D

E

F

G

H

I

J

K

L

M

N

O

P

R

S